Stochasticity and Quantum Chaos

Mathematics and Its Applications

Managing Editor:

M. HAZEWINKEL

Centre for Mathematics and Computer Science, Amsterdam, The Netherlands

Volume 317

Stochasticity and Quantum Chaos

Proceedings of the 3rd Max Born Symposium,
Sobótka Castle, September 15–17, 1993

edited by

Zbigniew Haba,
Wojciech Cegła
and
Lech Jakóbczyk

Institute of Theoretical Physics,
University of Wrocław,
Wrocław, Poland

SPRINGER-SCIENCE+BUSINESS MEDIA, B.V.

A C.I.P. Catalogue record for this book is available from the Library of Congress.

ISBN 978-0-7923-3230-5   ISBN 978-94-011-0169-1 (eBook)
DOI 10.1007/978-94-011-0169-1

Printed on acid-free paper

CONTENTS

FOREWORD

These are the proceedings of the Third Max Born Symposium which took place at Sobótka Castle in September 1993. The Symposium is organized annually by the Institute of Theoretical Physics of the University of Wrocław. Max Born was a student and later on an assistant at the University of Wrocław (Wrocław belonged to Germany at this time and was called Breslau). The topic of the Max Born Symposium varies each year reflecting the developement of theoretical physics. The subject of this Symposium "Stochasticity and quantum chaos" may well be considered as a continuation of the research interest of Max Born. Recall that Born treats his "Lectures on the mechanics of the atom" (published in 1925) as a first volume of a complete monograph (supposedly to be written by another person). His lectures concern the quantum mechanics of integrable systems. The quantum mechanics of non-integrable systems was the subject of the Third Max Born Symposium.

It is known that classical non-integrable Hamiltonian systems show a chaotic behaviour. On the other hand quantum systems bounded in space are quasiperiodic. We believe that quantum systems have a reasonable classical limit. It is not clear how to reconcile the seemingly regular behaviour of quantum systems with the possible chaotic properties of their classical counterparts. The quantum properties of classically chaotic systems constitute the main subject of these Proceedings. Other topics discussed are : the quantum mechanics of dissipative systems, quantum measurement theory, the role of noise in classical and quantum systems.

The Symposium came into being thanks to the enthusiasm of the lecturers and the substantial help of our colleagues. It has been supported by the Polish Research Council (KBN) and by the University of Wrocław. We are thankful to everybody who contributed to the success of this Symposium. Finally, we would like to thank Mrs. Anna Jadczyk for her excellent work in preparing the manuscripts.

Wrocław, February 28th 1994

The Editors

THE QUANTAL FATTENING OF FRACTALS

N.L. BALAZS
Department of Physics
State University of New York at Stony Brook
Stony Brook, NY 11794 - 3800

Abstract. Periodic points of a classical map may belong to a fractal set. It is shown here on a simple model that upon quantising this classical map the influence of the classical periodic points upon the quantal results is the same whether the periodic point belongs to a fractal set, or not. A simple intuitive explanation is given.

1. INTRODUCTION

How does one quantise fractals? Is there a meaning associated with these words?

Fractals were born through the *iteration* of functions of a complex variable giving rise to Julia sets (1918)[1]. They were thus intimately connected with the theory of (discrete) dynamical systems where the equations of dynamics are replaced by an *iterative* scheme, the mapping of the phase space onto itself. It is thus a natural step to generate fractal sets through the use of a dynamical system, and ask what happens to these fractal sets if we quantise the classical map.
A large amount of numerical evidence suggests that the classical periodic points (or orbits) have far reaching effects on the quantum mechanical results. In particular, we stress three such manifestations.
In the quantal description of a classical map the state of the system is specified by a vector in a finite dimensional vector space, and the time evolution is generated by a unitary operator acting on this space. This unitary operator, i.e. its eigenangles and eigenvalues contains all the information about the dynamical system.
The classical periodic points will leave their imprints on the quantal results in three different places.
a) There is strong numerical evidence that the presence of hyperbolic periodic points in the classical description changes the *statistics of the nearest neighbour distribution* of the eigenangles in a well defined manner[2];
b) the eigenvectors may be "scarred" by the periodic points[3];
c) in the coherent state representation the *correlations* between an initial state and its time evolved version should be large at those pq points which correspond to classical periodic points.
What happens, however, if the classical periodic points are fractals, hence form a *measure zero set*? This note will discuss this problem as exemplified on a simple model [4].

3

Z. Haba et al. (eds.), Stochasticity and Quantum Chaos, 3–11.

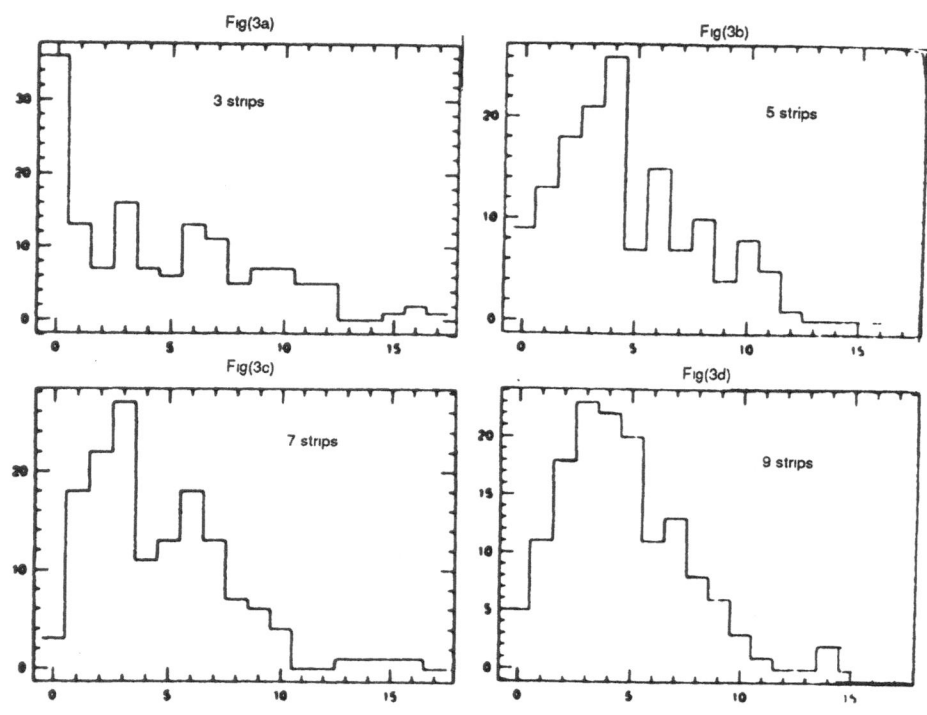

Fig. 1.   Nearest neighbour spacing distribution (nnsd) for quantum maps with different number of strips. The number of phase space cells, N, vary slightly, since they must be divisible by the number of strips. They are succesively 288, 290, 294, 288.

2. THE MODEL

Consider a classical discrete dynamical system that maps the unit square unto itself. The horizontal axis is labeled by q, the position (in units of the maximum displacement); the vertical axis is labeled by p, the momentum (in units of the maximum momentum). The transformation that represents the model is described geometrically as follows. Divide the unit square into three vertical strips. The first and third strips have the same width a. These vertical strips will be now changed into horizontal ones, and stacked on the top of each other. The first strip is made horizontal through a vertical compression of factor a, followed by a horizontal dilation of the same factor, becoming thereby a horizontal strip at the bottom. The second vertical strip is simply *rotated* into the horizontal position, becoming thereby the second horizontal strip. The third vertical strip is subjected to the same compression and dilation as the first, but laid down as the top horizontal strip. Thus, the evolution of the system is the iteration of the transformation

$$q' = q/a$$
$$p' = pa, \qquad \text{if} \quad 0 \le q \le a$$
$$q' = 1 - p$$
$$p' = q, \qquad \text{if} \quad a < q < 1 - a \qquad\qquad (1)$$
$$q' = (q - 1 + a)/a$$
$$p' = pa + 1 - a, \quad \text{if} \quad 1 - a \le q \le 1.$$

Take now, for simplicity, three strips of equal width, ie. a equal to one third. The original vertical central strip became horizontal. At the next iteration this horizontal central strip of width 1/3 looses the parts which overlap with the first and third vertical strip, due to their compression, dilation,and restacking. In fact, the lost pieces become horizontal strips, with a width reduced by the factor 1/3, blotting out the middle 1/9 strip of the top and bottom horizontal strips of width 1/3.These are regions to which some of the points of the original *rotating* central strip are dispersed. These are then regions where the dynamics will not become chaotic, where no hyperbolic points will reside. At the same time we notice that the construction described above corresponds to the blotting out operation used in the production of the *middle third Cantor set* ,$C_{1/3}$, on each axis. The blotted out segments result in a sequence of squares within which the dynamics is not chaotic. We note, however, that this construction blots out eventually the whole square leaving behind a fractal set of Hausdorff dimension 1 + log2/log3. Thus, the dynamics of this system results in the presence of periodic orbits of all periods, and a chaotic set of mesure zero!

A proper verification of this statement can immediately be given[4] if one constructs the symbolic dynamics of the transformation in which one gives the location of a point in the square by specifying p and q in a ternary basis, and forming a bi-infinite sequence out of them by writing qp back to back. The digits 0,1,2 represent the first, second and third vertical and horizontal partition of the square, and the dynamics can be easily translated into shifts and interchanges. From the symbolic dynamics one can also show that the periodic points present in the blotted out regions represent elliptic periodic orbits, inverse parabolic orbits and direct parabolic orbits, forming families imitating integrable systems. Furthermore, one can demonstrate that the special choice of setting the parameter a to be 1/3 does not alter these results since maps with different a values are topologically conjugate to each other. It is easy to generalise these maps . For example we may cut the unit square into 5 vertical strips of equal size, strip 3 being the central strip. Compress, stretch, stack the first and second onesin succession, as before. Then rotate the central strip into its horizontal position. Finally, compress, stretch, and stack the fourth and fifth strips as before.

3. THE QUANTISED MODEL

The quantum mechanics of the model displays the state of the system in a finite, N dimensional vector space, and the time evolution is described by a unitary operator that has all the classical symmetries, and possesses the correct classical limit. N is equal to the number of phase cells of the classical phase space of finite size; thus, in

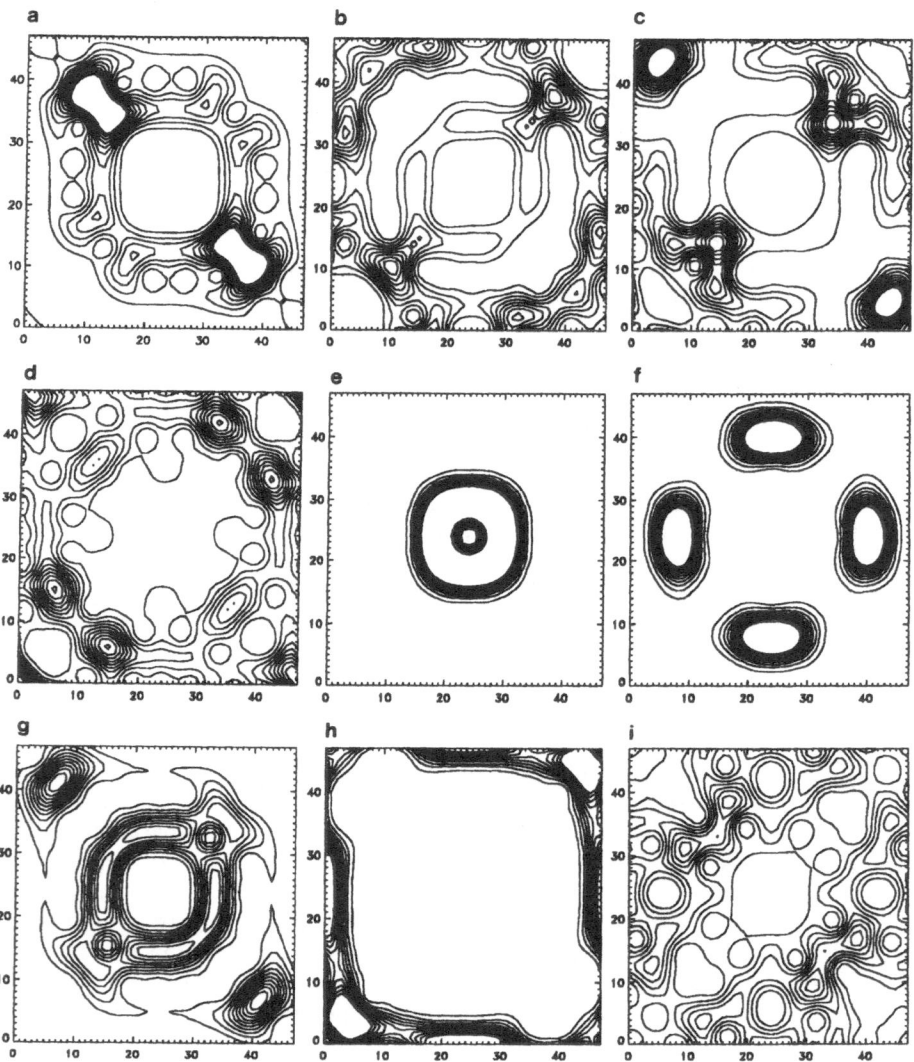

Fig. 2. Some eigenfunctions of the unitary propagator for the 3 strip map, with N = 48, exhibited in a coherent state representation. Nine eigenfunctions out of 48 are shown. For details see the text.

our reduced units Planck's constant is simply 1/N. In the present case the classical limit is simply $N \to \infty$, thus the number of phase cells go to infinity, while the size of the available phase space remains finite (contrary to the usual classical limit where the extent of the phase plane also tends to infinity).

The construction of the vector space and the unitary operator follows the prescription

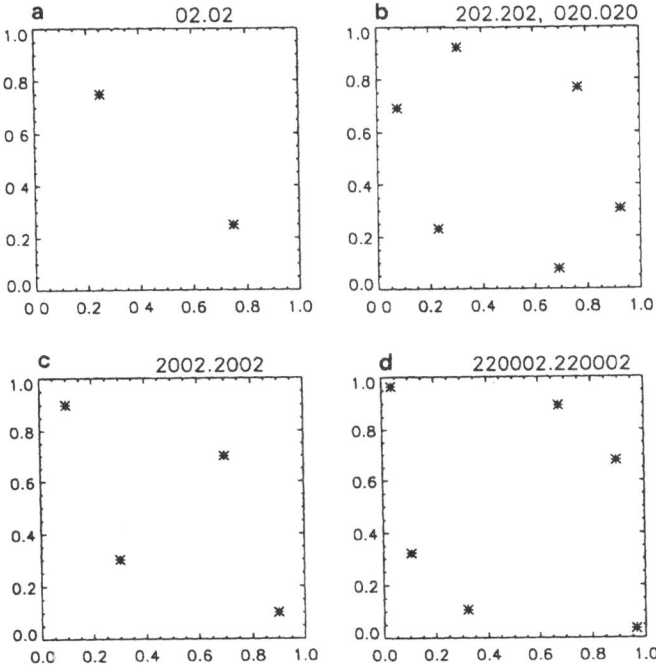

Fig. 3. Some periodic points (or orbits) of the 3 strip map that belong to the chaotic fractal set and scar some of the eigenfunctions shown in Fig.2.

given for the quantised Baker's Map[5], using the antiperiodic boundary conditions on the vector space, as introduced by Saraceno[6] to preserve fully the classical reflection symmetry around the center of the unit square. Accordingly, the discrete position eigenstates $|n>$, $n = 0,1,...,N-1$, and discrete momentum eigenstates $|m>$, $m = 0,1,...,N-1$, are repeated antiperiodically, e.g. $|n> = - |n+N>$. Either set of eigenvectors form a complete orthogonal basis in the vector space. The transformation matrix between the two bases is a NxN discrete Fourier transform matrix given by

$$(G_N)_{mn} = < m|n > = \frac{1}{\sqrt{N}} \exp[-2\pi \, i(m + 1/2)(n + 1/2)]. \tag{2}$$

The unitary one step propagator is given in the q representation by the matrix

$$B_1 = G_N^{-1} \begin{pmatrix} G_{Na} & 0 & 0 \\ 0 & I_{N(1-2a)} & 0 \\ 0 & 0 & G_{Na} \end{pmatrix}. \tag{3}$$

Here $I_{N(1-2a)}$ is the $N(1-2a)xN(1-2a)$ identity matrix. Since Na is required to be an integer, the cuts in the classical map are restricted to rational points.
The five equal strip map (mentioned before) will have the propagator

$$B_2 = G_N^{-1} \begin{pmatrix} G_{N/5} & 0 & 0 & 0 & 0 \\ 0 & G_{N/5} & 0 & 0 & 0 \\ 0 & 0 & I_{N/5} & 0 & 0 \\ 0 & 0 & 0 & G_{N/5} & 0 \\ 0 & 0 & 0 & 0 & G_{N/5} \end{pmatrix}.$$

4. RESULTS

Let us study now how the presence of the fractal set affects the eigenangle distribution, the eigenfunctions, and the correlations ?
a) Eigenangle distributions
The propagator is a unitary operator, hence its eigenvalues are specified by eigenangles. When a is equal 1/3 all eigenangles turn out to be irrational multiples of 2π, save the rational multiples of 0, $\pi/2$, π, and $3\pi/2$ which are associated with degenerate eigenvectors.
There is a great deal of empirical evidence that the nearest neighbour spacing (nns) distribution is Wigner like if the corresponding classical system is chaotic, and Poisson like if it is integrable[2]. The answers are less clear if the classical chaotic and integrable regions coexist. We expect, however, that the presence of mixed regions of finite measure will exhibit an intermediate nns distribution[2]. *What happens if the classical hyperbolic region is fractal?*

Consider the classical maps which are divided in 2k+1 equal strips: k on the left, k on the right. Their dynamics is similar to the three strip map, (k = 1). The central strip rotates, while the other strips undergo similar compression, stretching and stacking as before. The corresponding nns distributions are shown in Figure 1a - d, corresponding to k = 1,2,3,4. (The asssociated reduced Planck's constant varies slighty, since N/(2k+1) must be an integer.) The transition from Poisson like to intermediate statistics is unmistakeable. Since the Hausdorff dimension of the fractal set describing the classical hyperbolic orbit region is increasing with k this suggests that the measure zero set of chaotic orbits are indeed contributing to the quantal spectrum.

This is quite natural from an intuitive standpoint. Let us study the quantal description in a coherent state representation[7]. There we find that a *point* in the classical phase space corresponds to a *region* of size $2\pi\hbar$ in the mock phase-space displaying the coherent state representation. (In the present units this corresponds to a region of size 1/N). Thus, we expect that in this representation a fractal set in the classical phase space should correspond to a region of finite measure in the quantised version. *The fractal set becomes fattened.*
b) The eigenfunctions
It is fruitful to study the eigenfunctions in a coherent state representation, adapted

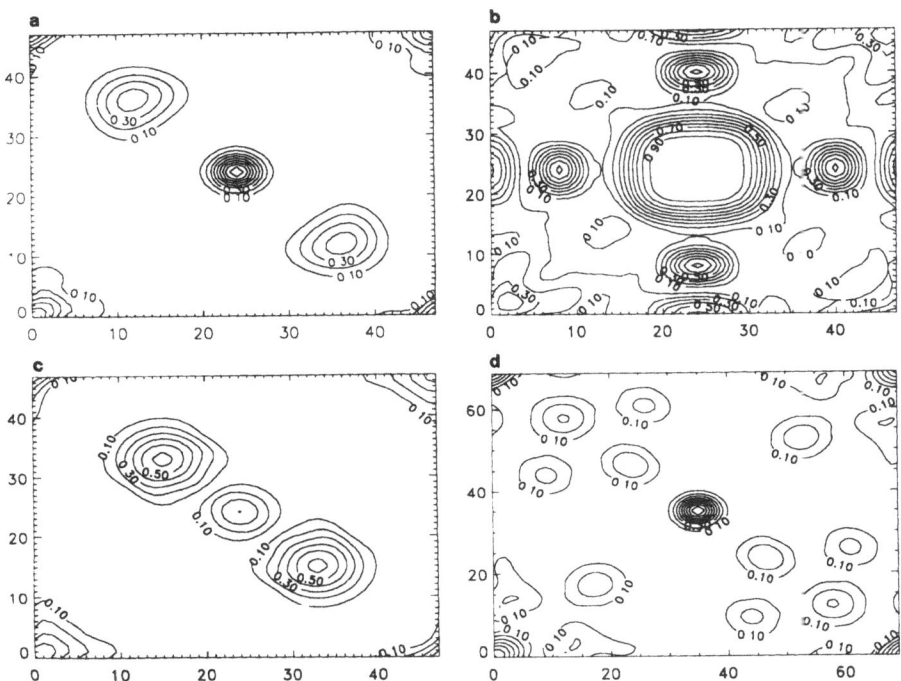

Fig. 4. Autocorrelation functions in a coherent state representation for some quantum maps, at different times n, and for different N values: a) 3 strip map, n = 2, N = 48; b) 3 strip map, n = 4, N = 48; asymmetric 3 strip map, first and last strip width 11/24, center strip width 2/24, n = 2, N = 48; d) five strip map, n = 2, N = 70.

to this class of problems[6]. The phenomenon of periodic orbit scarring has now been both observed[3] and understood[3a]. Their occurance in the present problem will be illustrated on some representative examples, culled from the eigenfunctions of the three strip map, with N = 48. They are shown as contour plots of the absolute value square in a coherent state representation. The majority of these functions are either concentrated around the regular parabolic and elliptic orbits; or, are scarred by periodic orbits belonging to the fractal set. For example, Fig. 2e shows an eigenstate supported by the trivial central region; Fig. 2f depicts an eigenstate associated with a period-8 parabolic island. On the other hand, Figs. 2a-d show eigenfunctions scarred by hyperbolic orbits out of the fractal set. The associated classical periodic orbits are shown on Fig. 3. Thus, the orbits in the fractal set perform their scarring the same way as orbits from a set of finite measure.

c) Autocorrelation functions

The quantity $<pq|B^n|pq>$ is the transition amplitude that a coherent state indexed

by p and q will recur after n steps. One expects that the absolute value of this amplitude should exhibit maxima at the corresponding classical periodic points of period n.

As an example consider the three strip map, for $n = 2$, $N = 48$. What are the classical periodic points of period two ? First of all , the center is an elliptic fixed point, hence it will be also a period two point. Similarly, the bottom left and top right corners are fixed points of the fractal hyperbolic set; hence they are also period two points. The points $(1/4, 3/4)$ and $(3/4, 1/4)$ are period two points of the fractal hyperbolic set. On Figure 4. we see that all these points are present in the correlation function; moreover, the points associated with the classical fractal set appear in the same way as the ones associated with the usual periodic points!

It is easy to comprehend why this be so. The classical fractal set was the outcome of a limiting process. Stopping before we reach the limit results in a set of finite measure. The correlation functions for a finite n are related to this finite set, obtained by arresting the blotting out operations after n steps.

One can show that the maxima of the autocorrelations over the Cantor set of hyperbolic points is increasing, while the contributions coming from the dominant parabolic and elliptic points have lower contributions. The interplay of these two features generates the shift in the nature of the spectral statistics.

CONCLUSION

On these simple examples we can see a dilution of the subtle measure-theoretical distinction between fractal sets of periodic orbits and sets of periodic orbits of finite measure as we diagnose the presence of these classical orbits through their imprints on the quantal description of the classical dynamical system.

ACKNOWLEDGMENTS

Most of these results are due to the work and insight of Dr. A. Lakshminarayan whose PH. D. Thesis contains not only these, but also many other results.

I am grateful to the organisers of the Symposium for their kind invitation, allowing me to pay in this manner my tribute to the memory of Max Born.

Part of this effort was supported by the National Science Foundation (USA)

References

1. K. J. FALCONER, "Fractal Geometry" Wiley, New York 1990
2. O. BOHIGAS AND M-J. GIANNONI, in "Mathematical and Computational Methods in Nuclear Physics" (J. S. Dehesa, J. M. G. Gomez, and A. Polls, Eds.) Lecture Notes in Physics, Vol.209. Springer-Verlag, New York/Berlin. 1984.
3. E. J. HELLER, Phys.Rev.Lett. 53 (1984) 515; a) M. V. BERRY, Proc Roy. Soc. London Ser. A 423 (1989) 219;
    E. B. BOGOMOLNY, Physica D 31 (1988) 169.
4. A. LAKSHMINARAYAN, Ph.D. Thesis, 1993, State University of New York at Stony Brook, Stony Brook N.Y.; Ann. Phys. (N.Y.) 225 (1993) 000.
5. N. L. BALAZS AND A. VOROS, Ann. Phys. (N.Y.) 190 (1989) 1.
6. M. SARACENO, Ann. Phys. (N.Y.) 199 (1990) 37.

HOW AND WHEN QUANTUM PHENOMENA BECOME REAL

PH. BLANCHARD
Faculty of Physics, University of Bielefeld
D-33615 Bielefeld

and

A. JADCZYK
Institute of Theoretical Physics. University of Wroclaw
Pl. Maxa Borna 9
PL-50204 Wroclaw

Abstract. We discuss recent developments in the foundations of quantum theory with a particular emphasis on description of measurement–like couplings between classical and quantum systems. The SQUID-tank coupling is described in some details, both in terms of the Liouville equation describing statistical ensembles and piecewise deterministic random process describing random behaviour of individual systems.

1. Introduction

Quantum theory is without doubt one of the most successful constructions in the history of theoretical physics and moreover the most powerful theory of physics: Its predictions have been perfectly verified until now again and again. The new conception of Nature proposed by Bohr, Heisenberg and Born was radically different from that of classical physics and several paradoxes have plagued Quantum Physics since its inception. Although the formalism of non relativistic quantum mechanics was constructed in the late 1920's the interpretation of Quantum Theory is still today the most controversial problem in the foundations of physics. The mathematical formalism and the orthodox interpretation of QM are stunningly simple but leave the gate open for alternative interpretations aimed at solving the dilemma lying in the Copenhagen interpretation. "The fact that an adequate presentation of QM has been so long delayed is no doubt caused by the fact that Niels Bohr brainwashed a whole generation of theorists into thinking that the job was done fifty years ago" wrote Murray Gell Mann 1979. It was also John Bell's point of view that "something is rotten" in the state of Denmark and that no formulation of orthodox quantum mechanics was free of fatal flows. This conviction motivated his last publication [1]. As he says "Surely after 62 years we should have an exact formulation of some serious part of quantum mechanics. By "exact" I do not mean of course "exactly true". I only mean that the theory should be fully formulated in mathematical terms, with nothing left to the discretion of the theoretical physicist ...". Orthodox Quantum Mechanics considers two types of incompatible time evolution U and R, U denoting the unitary evolution implied by Schrödinger's equation and R the reduction of the quantum state. U is linear, deterministic, local, continuous and time reversal invariant while R is probabilistic, non-linear, discontinuous and acausal. Two options are

Z. Haba et al. (eds.), Stochasticity and Quantum Chaos, 13–29.
© 1995 *Kluwer Academic Publishers.*

possible for completing Quantum Mechanics. According to John Bell [10] "Either the wave functions is not everything or it is not right ...".

In recent papers [2, 3, 4] we propose mathematically consistent models describing the information transfer between classical and quantum systems. The class of models we consider aims at providing an answer to the question of how and why quantum phenomena become real as a result of interaction between quantum and classical domains. Our results show that a simple dissipative time evolution can allow a dynamical exchange of information between classical and quantum levels of Nature. Indeterminism is an implicit part of classical physics and an explicit ingredient of quantum physics. Irreversible laws are fundamental and reversibility is an approximation. R. Haag formulated a similar thesis as "... once one accepts indetermination there is no reason against including irreversibility as part of the fundamental laws of Nature" [5]. According to the standard terminology the joint systems in our models are open. Thus one is tempted to try to understand their behaviour as an effective evolution of subsystems of unitarily evolving larger quantum systems. Although mathematically possible such an enlargement is non-unique. Therefore we prefer to extend the prevailing paradigm and learn as much as possible how to deal directly with open systems and incomplete information.

With a properly chosen initial state the quantum probabilities are exactly mirrored by the state of the classical system and moreover the state of the quantum subsystem converges as $t \to +\infty$ to a limit in agreement with von Neumann-Lüders standard quantum mechanical measurement projection postulate R. In our model the quantum system $\sum_q$ is coupled to a classical recording device $\sum_c$ which will respond to its actual state. $\sum_q$ should affect $\sum_c$, which should therefore be treated dynamically. We thus give a minimal mathematical semantics to describe the measurement process in Quantum Mechanics. For this reason the simplest models that we proposed can be seen as the elementary building blocks used by Nature in the constant communications that take place between the quantum and classical levels. In our framework a quantum mechanical measurement is nothing else as a coupling between a quantum mechanical system $\sum_q$ and a classical system $\sum_c$ via a completely positive semigroup $\alpha_t = e^{tL}$ in such a way that information can be transferred from $\sum_q$ to $\sum_c$. A measurement represents an exchange of information between physical systems and therefore involves entropy production. See [6] where a definition of entropy for non-commutative systems is given, which is based on the concepts of conditional entropy and stationary couplings between $\sum_q$ and $\sum_c$. Moreover Sauvageot and Thouvenot show the equivalence of this definition with the one proposed by Connes, Narnhofer and Thirring [12].

There have been many attempts to explain quantum measurements. For recent reviews see [7, 8, 9]. Our claim is, that whatever the mechanism used to derive models of measurements starting from fundamental interactions is, this mechanism will lead finally to one model of the class we introduced. In fact any realistic situation will reduce to a model of our class since the overwhelming majority of the properties of the counter and the environment are irrelevant from the point of view of statistically predicting the result of a measurement. We propose indeed to consider the total system $\sum_{tot} = \sum_q \otimes \sum_c$, and the behaviour associated to the total algebra of observables $\mathcal{A}_{tot} = \mathcal{A}_q \otimes \mathcal{A}_c = \mathcal{C}(X_c) \otimes \mathcal{L}(\mathcal{H}_q)$, where X_c is the classical phase space and $\mathcal{H}_q$ the Hilbert space associated to $\sum_q$, is now taken as the fundamental reality with pure quantum behaviour as an approximation valid in the cases when recording effects can be neglected. In $\mathcal{A}_{tot}$ we can describe irreversible changes occuring in the physical world, like the blackening photographic emulsion, as well as idealized reversible pure quantum and pure classical processes. We extend the model of Quantum Theory in such a way that the successful features of the exisiting theory are retained but the transitions between equilibria in the sense of recording effects are permitted. In Section 2 we will briefly describe the mathematical and physical ingredients of the simplest model and discuss the measurement process in this framework.

The range of applications of the model is rather wide as will be shown in Section 3 with a discussion of Zeno effect, giving an account of [11]. To the Liouville equation describing the time evolution of statistical states of $\sum_{tot}$ we will be in position to associate a piecewise deterministic process taking values in the set of pure states of $\sum_{tot}$. Knowing this process one can answer all kinds of questions about time correlations of the events as well as simulate numerically the possible histories of individual quantum-classical systems. Let us emphasize that nothing more can be expected from a theory without introducing some explicit dynamics of hidden variables. What we achieved is the maximum of what can be achieved, which is more than orthodox interpretation gives. There are also no paradoxes; we cannot predict individual events (as they are random), but we can simulate the observations of individual systems. Moreover, we will briefly comment on the meaning of the wave function. The purpose of Section 4 is to discuss the coupling between a SQUID and a damped classical oscillating circuit. Section 5 deals there with some concluding remarks.

The support of the Polish KBN and German Alexander von Humboldt-Foundation is acknowledged with thanks.

2. Communicating classical and quantum systems

Measurements provide the link between theory and experiment and their analysis is therefore one of the most important and sensitive parts of any interpretation.

For a long time the theory of measurements in quantum mechanics, elaborated by Bohr, Heisenberg und von Neumann in the 1930s has been considered as an esoteric subject of little relevance for real physics. But in the 1980s the technology has made possible to transform "Gedankenexperimente" of the 1930s into real experiments. This progress implies that the measurement process in quantum theory is now a central tool for physicists testing experimentally by high-sensitivity measuring devices the more esoteric aspects of Quantum Theory.

Quantum mechanical measurement brings together a macroscopic and a quantum system.

Let us briefly describe the mathematical framework we will use. A good deal more can be said and we refer the reader to [2, 3, 4]. Our aim is to describe a non-trivial transfer of information between a quantum system $\sum_q$ in interaction with a classical system $\sum_c$. To the quantum system there corresponds a Hilbert space $\mathcal{H}_q$. In $\mathcal{H}_q$ we consider a family of orthonormal projectors $e_i = e_i^* = e_i^2$, $(i = 1, ..., n)$, $\sum_{i=1}^n e_i = 1$, associated to an observable $A = \sum_{i=1}^{n} \lambda_i e_i$ of the quantum mechanical system. The classical system is supposed to have m distinct pure states, and it is convenient to take $m \geq n$. The algebra $\mathcal{A}_c$ of classical observables is in this case nothing else as $\mathcal{A}_c = \mathbf{C}^m$. The set of classical states coincides with the space of probability measures. Using the notation $X_c = \{s_0, ..., s_{m-1}\}$, a classical state is therefore an m-tuple $p = (p_0, ..., p_{m-1}), p_\alpha \geq 0, \sum_{\alpha=0}^{m-1} p_\alpha = 1$. The state s_0 plays in some cases a distinguished role and can be viewed as the neutral initial state of a counter. The algebra of observables of the total system $\mathcal{A}_{tot}$ is given by

$$\mathcal{A}_{tot} = \mathcal{A}_c \otimes L(\mathcal{H}_q) = \mathbf{C}^m \otimes L(\mathcal{H}_q) = \bigoplus_{\alpha=0}^{m-1} L(\mathcal{H}_q), \tag{1}$$

and it is convenient to realize $\mathcal{A}_{tot}$ as an algebra of operators on an auxiliary Hilbert space $\mathcal{H}_{tot} = \mathcal{H}_q \otimes \mathbf{C}^m = \bigoplus_{\alpha=0}^{m-1} \mathcal{H}_q$. $\mathcal{A}_{tot}$ is then isomorphic to the algebra of block diagonal $m \times m$ matrices $A = diag(a_0, a_1, ..., a_{m-1})$ with $a_\alpha \in L(\mathcal{H}_q)$. States on $\mathcal{A}_{tot}$ are represented by block diagonal matrices

$$\rho = diag(\rho_0, \rho_1, ..., \rho_{m-1}) \tag{2}$$

where the ρ_α are positive trace class operators in $L(\mathcal{H}_q)$ satisfying moreover $\sum_\alpha Tr(\rho_\alpha) = 1$. By taking partial traces each state ρ projects on a 'quantum state' $\pi_q(\rho)$ and a 'classical state' $\pi_c(\rho)$ given respectively by

$$\pi_q(\rho) = \sum_\alpha \rho_\alpha, \tag{3}$$

$$\pi_c(\rho) = (Tr\rho_0, Tr\rho_1, ..., Tr\rho_{m-1}). \tag{4}$$

The time evolution of the total system is given by a semi group $\alpha^t = e^{tL}$ of positive maps[1] of $\mathcal{A}_{tot}$– preserving hermiticity, identity and positivity – with L of the form

$$L(A) = i[H, A] + \sum_{i=1}^{n} (V_i^* A V_i - \frac{1}{2}\{V_i^* V_i, A\}). \tag{5}$$

The V_i can be arbitrary linear operators in $L(\mathcal{H}_{tot})$ such that $\sum V_i^* V_i \in \mathcal{A}_{tot}$ and $\sum V_i^* A V_i \in \mathcal{A}_{tot}$ whenever $A \in \mathcal{A}_{tot}$, H is an arbitrary block-diagonal self adjoint operator $H = diag(H_\alpha)$ in $\mathcal{H}_{tot}$ and $\{,\}$ denotes anticommutator i.e.

$$\{A, B\} \equiv AB + BA. \tag{6}$$

In order to couple the given quantum observable $A = \sum_{i=1}^{n} \lambda_i e_i$ to the classical system, the V_i are chosen as tensor products $V_i = \sqrt{\kappa} e_i \otimes \phi_i$, where ϕ_i act as transformations on classical (pure) states. Denoting $\rho(t) = \alpha_t(\rho(0))$, the time evolution of the states is given by the dual Liouville equation

$$\dot{\rho}(t) = -i[H, \rho(t)] + \sum_{i=1}^{n} (V_i \rho(t) V_i^* - \frac{1}{2}\{V_i^* V_i, \rho(t)\}), \tag{7}$$

---

[1] In fact, the maps we use happen to be also completely positive.

where in general H and the V_i can explicitly depend on time.

Remarks:

1) It is possible to generalize this framework for the case where the quantum mechanical observable A we consider has a continuous spectrum (as for instance in a measurement of the position) with $A = \int_{\mathbf{R}} \lambda dE(\lambda)$. See [13, 14] for more details. It is also straightforward to include simultaneous measurements of noncommuting observables via semi–spectral measures (see [15]).

2) Since the center of the total algebra $\mathcal{A}_{tot}$ is invariant under any automorphic unitary time evolution, the Hamiltonian part H of the Liouville operator is not directly involved in the process of transfer of information from the quantum subsystem to the classical one. Only the dissipative part can achieve such a transfer in a finite time.

In [2] we propose a simple, purely dissipative Liouville operator (i.e. we put $H = 0$) that describes an interaction of $\sum_q$ and $\sum_c$, for which $m = n + 1$ and $V_i = e_i \otimes \phi_i$, where ϕ_i is the flip transformation of X_c transposing the neutral state s_0 with s_i. We show that the Liouville equation can be solved explicitly for any initial state $\rho(0)$ of the total system. Assume now that we are able to prepare at time $t = 0$ the initial state of the total system Σ_{tot} as an uncorrelated product state $\rho(0) = w \otimes P^\epsilon(0)$, $P^\epsilon(0) = (p_0^\epsilon, p_1^\epsilon, ..., p_n^\epsilon)$ as initial state of the classical system parametrized by $\epsilon, 0 \leq \epsilon \leq 1$:

$$p_0^\epsilon = 1 - \frac{n\epsilon}{n+1}, \tag{8}$$

$$p_1^\epsilon = \frac{\epsilon}{n+1}. \tag{9}$$

In other words for $\epsilon = 0$ the classical system starts from the pure state $P(0) = (1, 0, ..., 0)$ while for $\epsilon = 1$ it starts from the state $P'(0) = (\frac{1}{n+1}, \frac{1}{n+1}, ..., \frac{1}{n+1})$ of maximal entropy. Computing $p_i(t) = T^-(\rho_i(t))$ and then the normalized distribution

$$\tilde{p}_i(t) = \frac{p_i(t)}{\sum_{r=1}^n p_r(t)} \tag{10}$$

with $\rho(t) = (\rho_0(t), \rho_1(t), ..., \rho_n(t))$ the state of the total system we get:

$$\tilde{p}_i(t) = q_i + \frac{\epsilon(1 - nq_i)}{\epsilon n + \frac{(1-\epsilon)(n+1)}{2}(1 - e^{-2\kappa t})}, \tag{11}$$

where we introduced the notation

$$q_i = Tr(e_i w), \tag{12}$$

for the initial quantum probabilities to be measured. For $\epsilon = 0$ we have $\tilde{p}_i(t) = q_i$ for all $t > 0$, which means that the quantum probabilities are exactly, and immediately after switching on of the interaction, mirrored by the state of the classical system. For $\epsilon = 1$ we get $\tilde{p}_i(t) = 1/n$. The projected classical state is still the state of maximal entropy and in this case we get no information at all about the quantum state by recording the time evolution of the classical one. In the intermediate regime, for $0 < \epsilon < 1$, it is easy to show that $|\tilde{p}_i(t) - q_i|$ decreases at least as $2\epsilon(1 + e^{-2\kappa t})$ with $\epsilon \to 0$ and $t \to +\infty$. For $\epsilon = 0$, that is when the measurement is exact, we get for the partial quantum state

$$\pi_q(\rho(t)) = \sum_i e_i w e_i + e^{-\kappa t}(w - \sum_i e_i w e_i),$$

so that

$$\pi_q(\rho(\infty)) = \sum_i e_i w e_i, \tag{13}$$

which means that the partial state of the quantum subsystem $\pi_q(\rho(t))$ tends for $\kappa t \to \infty$ to a limit which coincides with the standard von Neumann-Lüders quantum measurement projection postulate.

Remark:

The normalized distribution $\tilde{p}_i(t)$ is nothing else as the read off from the outputs $s_1 ... s_n$ of the classical system $\sum_c$.

To discuss now the interplay between efficiency and accuracy by measurement, let us consider the case where

$$V_i = \sqrt{\kappa} e_i \otimes f_i, \tag{14}$$

f_i being the transformation of X_c mapping s_0 into s_i. In the Liouville equation we consider also an Hamiltonian part. We find for the Liouville equation:

$$\begin{aligned} \dot{\rho}_0 &= -i[H, \rho_0] - \kappa \rho_0, \\ \dot{\rho}_i &= -i[H, \rho_i] + \kappa e_i \rho_0 e_i, \end{aligned} \tag{15}$$

where we allow for time dependence i.e. $H = H(t), e_i = e_i(t)$. Setting $r_0(t) = Tr(\rho_0(t))$, $r_i(t) = Tr(\rho_i(t))$, and assuming that the initial state is of the form $\rho = (\rho_0, 0, ..., 0)$ we conclude that $\dot{r}_0 = -\kappa r_0$ and thus $r_0(t) = e^{-\kappa t}$ which obviously implies that

$$\sum_{i=1}^{n} r_i(t) = 1 - e^{-\kappa t}, \tag{16}$$

from which it follows that a 50 % efficiency requires $\log 2/\kappa$ time of recording. It is easy to compute $r_i(t)$ and

$$\tilde{p}_i(t) = \frac{r_i(t)}{\sum_{j=1}^{n} r_j(t)}$$

for small t. We get

$$\tilde{p}_i(t) = q_i + \frac{\kappa^2 t^2}{2} \frac{1}{\kappa} \langle \frac{de_i}{dt} \rangle_{\rho_0} + o(t^2), \tag{17}$$

where

$$\frac{de_i}{dt} \doteq \frac{\partial e_i}{\partial t} + i[H, e_i]. \tag{18}$$

Efficiency requires $\kappa t \gg 1$ while accuracy is achieved if $(\kappa t)^2 \ll \frac{\kappa}{\langle \dot{e}_i \rangle_{\rho_0}}$. To monitor effectively and accurately fast processes we must therefore take $\kappa \ll 10^2 \langle \dot{e}_i \rangle_{\rho_0}$. Suppose now that H and e_i does not depend on time. Then it is easy to show that if either $\rho_0(0)$ or e_i commutes with H, we get $\tilde{p}_i(t) = q_i$ exactly and instantly.

In [3, 4] we describe and analyze a Stern Gerlach experiment and a model of a counter for a one-dimensional ultra-relativistic quantum mechanical particle.

3. Quantum Zeno Effect

Zeno of Elea is famous for the paradoxes whereby, in order to recommend the doctrine of the existence of "the one" (i.e. indivisible reality) he sought to controvert the common-sense belief in the existence of "the many" (i.e. distingnishable quantities and things capable of motion). The quantum Zeno effect was described many years ago when it was claimed that is possible to inhibit or even to stop the decay of an unstable quantum mechanical system by performing a sequence of frequent measurements. The exponential decay law $P(t) = e^{-\gamma t}$ is experimentally confirmed for most unstable particles and nuclei in a wide range of time. The initial decay rate is in this case $-\frac{dP}{dt}(0_+) = \gamma$. On the other hand, from quantum theory, we get for the decay law $\dot{P}_\psi(0) = 2Im < \psi, H\psi >= 0$ with $P_\psi(t) = | < \psi, e^{-itH}\psi > |^2$. When the particle is observed at $t/n, 2t/n, \ldots$ then $P_\psi(t) = P_\psi(t/n)^n$; now if $\dot{P}_\psi(0_+) = 0$ if follows that $\lim_{n \to +\infty} P_\psi(t/n)^n = 1$, which implies that frequent observations freeze the system in its initial state.

Using our model of a continuous measurement we can easily discuss this effect for a quantum spin $1/2$ system coupled to a 2-state classical system [11]. We consider only one orthogonal projector e on the Hilbert space $\mathcal{H}_q = \mathbf{C}^2$. To specify the coupling dynamics we choose the coupling operator V in the following symmetric way:

$$X = \sqrt{\kappa} \begin{pmatrix} 0, & e \\ e, & 0 \end{pmatrix}. \tag{19}$$

The Liouville equation for the total state $\rho = diag(\rho_0, \rho_1)$ reads now

$$\dot{\rho}_0 = -i[H, \rho_0] + \kappa(e\rho_1 e - \tfrac{1}{2}\{e, \rho_0\}), \tag{20}$$

$$\dot{\rho}_1 = -i[H, \rho_1] + \kappa(e\rho_0 e - \tfrac{1}{2}\{e, \rho_1\}). \tag{21}$$

The partial quantum state $\pi_q(\rho) = \hat{\rho} = \rho_0(t) + \rho_1(t)$ evolves in this particular model independently of the state of the classical system, which expresses the fact that we have here only transport of information from $\sum_q$ to $\sum_c$. The time evolution of $\hat{\rho}(t)$ is given by

$$\dot{\hat{\rho}} = -i[H, \hat{\rho}] + \kappa(e\hat{\rho}e - \frac{1}{2}\{e, \hat{\rho}\}). \tag{22}$$

Let us now choose the Hamiltonian part $H = \frac{\omega}{2}\sigma_3$, and $e = \frac{1}{2}(\sigma_0 + \sigma_1)$, and start with the quantum system $\sum_q$ being for $t = 0$ in the eigenstate of σ_1. We repeatedly check with "frequency" κ if the system is still in this initial state, each "yes" inducing a flip in the coupled classical device, which we continuously observe. The solution of (23) such that $\hat{\rho}(0) = e$ can be easily found. Moreover it is possible for strongly coupled system i.e. for $\kappa t >> 1$ and $\kappa/\omega >> 1$ to obtain asymptotic formulae for the distance travelled by the quantum state $d(\hat{\rho}(t), e)$ in the Bures or in the Frobenius norm $\| \hat{\rho} \|^2 = Tr(\hat{\rho}^2)$. In this asymptotic regime we can show that the Bures distance achieved during the coupling is given by

$$d(\hat{\rho}(t), e) \approx \omega\sqrt{t/\kappa} \qquad (23)$$

The effect of slowing down the evolution of the quantum system can be confirmed by an independent, strong but non-demolishing, coupling of a third classical device. In [4, 13] we show moreover that a piecewise deterministic Markov process taking values on pure states of the total system is naturally associated to the Liouville equation and that the coupling constant κ is the average frequency of jumps of the classical system between its two states.

Remarks on "meaning of the wave function" It is tempting to use the Zeno effect for slowing down the time evolution in such a way, that the state of a quantum system Σ_q can be determined by carrying out measurements of sufficiently many observables. This idea, however, would not work, similarly like would not work the proposal of "protective measurements" of Y. Aharonov et al (see [16] [17]). To apply Zeno-type measurements just as to apply a "protective measurement" one would have to know the state beforehand. Our results suggest that obtaining a reliable knowledge of the quantum state may necessarily lead to a significant, irreversible disturbance of the state. This negative statement does not mean that we have shown that the quantum state cannot be objectively determined. We believe however that dynamical, statistical and information-theoretical aspects of the important problem of obtaining a "*maximal reliable knowledge* ;, *of the unknown quantum state with a least possible disturbance*" are not yet sufficiently understood.

4. SQUID - Tank circuit interaction

Superconductivity was discovered 1911 by Kamerlingh Onnes. Two important properties of superconductors set them apart from normal metallic conductors; they exhibit zero electrical resistance to current flow and they expel magnetic fields (the Meissner effect). In addition superconductors display a special characteristic when two are coupled through a thin insulating layer (the Josephson effect). Josephson devices consist of two superconducting films through which electrons can tunnel from one superconductor to the other. The tunneling can be by superconducting pairs via the Josephson effect. In a Josephson junction the current I is given by $I = I_c \sin \phi$, where I_c is the dissipationless current that the junction will sustain. The Josephson energy E, which is the kinetic energy of the current I flowing through the junction is given by $E = -\frac{\hbar}{2e}I_c \cos \phi$ and ϕ is the quantum mechanical phase difference across the function.

A SQUID is a ring-shaped superconducting circuit containing one or more so called weak links whose behaviour is governed by the Josephson equations of super-conductivity. A magnetic field applied to a SQUID alters its electrical characteristics. The ring's response can be interrogated with conventional electronics. SQUIDS posses a wide variety of macroscopic quantum mechanical properties. In recent years these has been considerable discussion of the dynamics of a system consisting of a SQUID coupled to a dissipative classical linear oscillator [18, 19, 20, 21]. Our aim is to show that a continuous version of our framework is very well adapted to discuss the behaviour of the coupled system consisting of a macroscopic classical system (tank circuit) and a single macroscopic quantum object (SQUID).

4.1. SQUID COUPLED TO A DAMPED CLASSICAL OPERATOR

A SQUID (Superconducting Quantum Interference Device) consists of a piece of superconductor with two holes that nearly connect at the "weak link". Suppose now that the device is in a state where a current is flowing round one of the holes and induces therefore a magnetic field whose field lines pass through this hole. The magnetic flux in such a ring is quantized, where the "flux quantum" is given by $\frac{h}{2e}$. Under the assumption that the circulating current is very small there is only one flux quantum say in the left hole. A macroscopically distinct state would be the symmetric case where one flux quantum is localized in the right hole. Quantum theory says that a SQUID can exits also in a state where the flux is delocalized between the two holes; flux quanta can pass from one hole to another by quantum tunnelling processes. SQUIDS are laboratory versions of Schrödinger's cat. The flux Φ trapped through the ring is a macroscopic variable which obeys a standard Schrödinger equation with the mass M replaced by the capacitance of the Josephson junction and the potential $V(\Phi)$ such that $\lim_{|\Phi| \to \infty} V(\Phi) = +\infty$. Our aim is to describe the interaction between a SQUID and a classical damped oscillator. Both are macroscopic electromagnetic circuits. The classical system can be seen as a model for a local environment for the SQUID. The Hamiltonian of the quantum system contains a source term through wich it can be coupled to the classical device.

The radiofrequency (rf) SQUID is a superconducting loop which is interrupted by a thin insulating layer (Josephson tunnel junction). The conduction electrons in the superconductor are paired, and thus encounter negligible dissipation within the body of the superconductor. The Cooper pairs tunnel through the insulating barrier. In general there is some dissipation associated with this process, as well as a capacitance determined by the geometry of the junction. A superconducting screening current I_S flows around the SQUID loop inductance L in response to an externally applied magnetic flux ϕ_{ext} generated by a magnetic field orthogonal to the SQUID for suitably chosen device parameter, the net flux obeys an equation of motion similar to that of a particle moving in a double well potential, with the capacitance C and conductance $1/R$ corresponding to the particle mass and dissipation

$$C\ddot{\Phi} + \frac{1}{R}\dot{\Phi} = -\frac{\partial V}{\partial \Phi}. \tag{24}$$

The potential V is given as a function of the net flux and the external flux Φ_{ext}

$$V(\Phi) = \frac{1}{2\Lambda}(\Phi - \phi_{ext})^2 - \frac{I_0 \Phi_0}{2\pi} \cos 2\pi \frac{\Phi}{\Phi_0}. \tag{25}$$

At a flux of $\Phi_0/2$ the screening is equal in magnitude in both wells, but opposite in direction. This is the optimal point for the observation of coherent tunneling effects. The net flux Φ may be considered to be a macroscopic variable since the condensate of superconducting pairs is described by the product of a large number of pair wave functions. The large overlap of the wave functions of the Cooper pairs, which are single quantum states extending over a macroscopic distance, produces a macroscopic phase coherence. SQUIDS are devices exhibiting quantized flux but being describable by classical quantities such as voltage and current. The total magnetic flux Φ is conjugate to the total electric flux i.e. the charge Q across the weak link. Then satisfy the commutation relation

$$[\Phi, Q] = i\hbar \mathbf{1}. \tag{26}$$

The Hamilton operator for the SQUID-tank model is given by

$$H(\varphi) = \frac{Q^2}{2C} + \frac{(\Phi - \phi_{ext})^2}{2\Lambda} - \hbar\omega \cos\left(\frac{2\pi\Phi}{\Phi_0}\right) \tag{27}$$

where

$$\phi_{ext} = \varphi_{ext} + \mu\varphi \tag{28}$$

$$Q = -i\hbar\frac{d}{d\Phi} \tag{29}$$

$$\Phi_0 = \frac{h}{2e}. \tag{30}$$

For the tank equation following Spiller et al. [19] one obtains

$$\ddot{\varphi} + \frac{\dot{\varphi}}{R_t C_t} + \frac{\varphi}{L_t C_t} = \frac{1}{C_t}\left(I_{\mathbf{IN}}(t) + \mu < \frac{\Phi - \phi_{ext}}{\Lambda} >\right). \tag{31}$$

States of the classical system are probabilistic measures p on its phase space $\Omega = (\mathbf{R}^2, d\varphi d\pi)$; we take for the canonical variables: the magnetic flux φ and its rate of change π (thought of as $\dot{\varphi}$). Because of dumping in the tank circuit, the equation of motion for φ cannot be written in a standard Hamiltonian form (but it can be written using a complex Hamiltonian – see e.g. [22]). For our purpose we need only the Liouville equation – which is just continuity equation for the classical flow:

$$\frac{\partial p}{\partial t} + \frac{\partial}{\partial\varphi}(\dot{\varphi}p) + \frac{\partial}{\partial\pi}(p\dot{\pi}) = 0. \tag{32}$$

Denoting $I_S := < \frac{\Phi - \phi_{ext}}{\Lambda} >$, with $\dot{\varphi} \to \pi$, $\dot{\pi} \to \ddot{\varphi}$ we obtain

$$\frac{\partial p}{\partial t} + \frac{\partial}{\partial\varphi}(\pi p) + \frac{\partial}{\partial\pi}\left(-p\left(\frac{\pi}{R_t C_t} + \frac{\varphi}{L_t C_t} + \frac{1}{C_t}\left(I_{\mathbf{N}} + \mu < I_S >\right)\right)\right) \tag{33}$$

or $\dot{p} = L_{cl}p$ where

$$L_{cl}p = -\pi\frac{\partial p}{\partial\varphi} + \left\{\frac{1}{R_t C_t}\pi + \frac{1}{L_t C_t}\varphi - \frac{1}{C_t}(I_{\mathbf{IN}} + \mu < I_S >)\right\} \times$$
$$\times \frac{\partial p}{\partial\pi} + \frac{1}{R_t C_t}p. \tag{34}$$

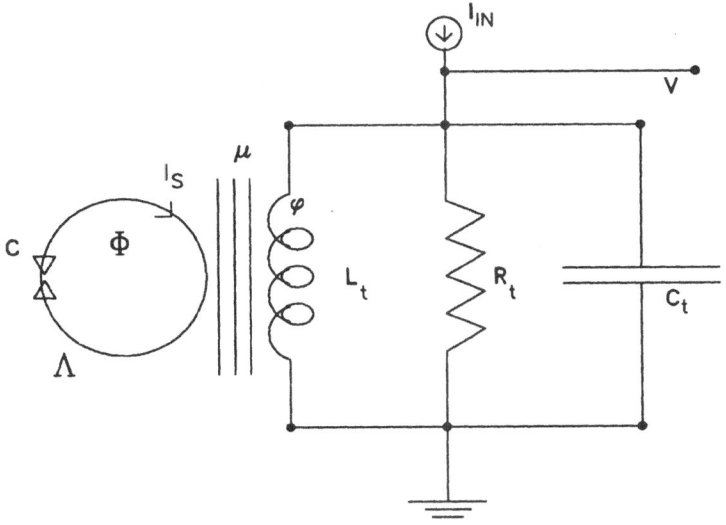

Fig. 1.

States of the total system consisting of SQUID and tank are described by measures $\rho(\varphi, \pi)$ on Ω with values in positive trace class operators in $\mathcal{H}_q = L^2(\mathbf{R}, d\Phi)$, normalized by

$$\int Tr(\rho(\varphi, \pi)) d\varphi \, d\pi = 1 \ . \tag{35}$$

It is convenient in the spirit of Section 2 to introduce the Hilbert space $\mathcal{H}_{tot}$ for the total system:

$$\mathcal{H}_{tot} = \int^{\oplus} \mathcal{H}_q \, d\varphi \, d\pi. \tag{36}$$

The vector state of $\mathcal{H}_{tot}$ are then given by functions $\Psi : (\varphi, \pi) \longmapsto \Psi(\varphi, \pi) \in \mathcal{H}_q$.

The coupling between SQUID and tank is postulated to be given by the following operator (generalizing obviously the Lindblad form to a continuous family)

$$L_{int} \rho = \int_{-\infty}^{+\infty} da(V_a \rho V_a^* - \frac{1}{2}\{V_a^* V_a, \rho\}) \tag{37}$$

with V_a given by

$$(V_a \Psi)(\varphi, \pi) = f(\Phi - \phi_{ext} - a)\Psi(\varphi, \pi - ka) \ . \tag{38}$$

The function f can be thought of as defining a sensitivity window - it should be odd or even:

$$f(x) = \pm f(-x) \ . \tag{39}$$

We denote

$$\alpha = \int_{-\infty}^{+\infty} f^2(x)dx \ . \tag{40}$$

The constant k (having dimension $[time]^{-1}$) is the second constant characterizing the coupling. We first notice that

$$\int_{-\infty}^{+\infty} V_a^* V_a \, da = \alpha \cdot I \tag{41}$$

so that

$$(L_{int}\rho)(\varphi, \pi) = \int da f(\Phi - \phi_{ext} - a)\rho(\varphi, \pi - ka)f(\Phi - \phi_{ext} - a)$$
$$-\alpha\rho(\varphi, \pi) \ . \tag{42}$$

The Liouville operator for the total system is then given by a sum of three terms

$$\begin{aligned} (\mathcal{L}\rho)(\varphi, \pi) = &-i[H(\varphi), \rho(\varphi, \pi)] + \\ &+ (L_{cl}\rho)(\varphi, \pi) + \\ &+ (L_{int}\rho)(\varphi, \pi) \ . \end{aligned} \tag{43}$$

In the following we will denote by $< F >$ the average for a quantity F:

$$< F >= \int Tr(F(\varphi, \pi)\rho(\varphi, \pi))d\varphi \, d\pi \ . \tag{44}$$

Therefore the time derivative of averages is given explicitly by

$$\begin{aligned} < \dot{F} > &= \int Tr(F(\varphi, \pi)\dot{\rho}(\varphi, \pi))d\varphi \, d\pi = \\ &= \int Tr(F(\varphi, \pi)(\mathcal{L}\rho)(\varphi, \pi))d\varphi \, d\pi \ . \end{aligned} \tag{45}$$

Let us compute

$$< \dot{\varphi} >= \int \varphi Tr((\mathcal{L}\rho)(\varphi, \pi))d\varphi \, d\pi \ . \tag{46}$$

Only the classical part contributes and we get

$$< \dot{\varphi} >=< \pi > \ . \tag{47}$$

We need also to compute $< \ddot{\varphi} >=< \dot{\pi} >$

$$< \dot{\pi} >= \int \pi Tr(\mathcal{L}\rho(\varphi, \pi))d\varphi \, d\pi \ . \tag{48}$$

The quantum Hamiltonian does not contribute while the classical part gives nothing else as the RHS of the classical equations of motion:

$$-\frac{<\dot{\varphi}>}{R_t C_t} - \frac{<\varphi>}{L_t C_t} + \frac{1}{C_t}I_{IN}(t) \ . \tag{49}$$

We compute next the term coming from L_{int}:

$$\int \pi Tr(f^2(\Phi - \phi_{ext} - a)\rho(\varphi, \pi - ka)da\, d\varphi\, d\pi$$

$$-\alpha \int \pi Tr(\rho(\varphi, \pi))d\varphi\, d\pi \quad . \tag{50}$$

Let us consider the first term. Changing variables $\pi - ka = \pi'$ we get

$$\int (\pi' + ka)Tr(f^2(\Phi - \phi_{ext} - a)\rho(\varphi, \pi'))da\, d\varphi\, d\pi' \, . \tag{51}$$

The π' term cancels with the last term in (50). We change also the a variables introducing a' by

$$a - \Phi - \phi_{ext} = a' \tag{52}$$

and obtain just

$$k \int Tr((a' + \Phi - \phi_{ext})f^2(a')\rho(\varphi, \pi))da'\, d\varphi\, d\pi \, . \tag{53}$$

The term $\int a' f^2(a')$ gives zero because f^2 is assumed to be even. What remains is

$$\alpha k \int Tr((\Phi - \phi_{ext})\rho(\varphi, \pi))d\varphi\, d\pi =$$

$$= \alpha k < \Phi - \phi_{ext} > \, . \tag{54}$$

It follows that our evolution law for averages is compatible with that of Spiller [18] if we put

$$\alpha k = \frac{\mu}{C_t \Lambda} \, , \tag{55}$$

which fixes one of the parameters α, k in terms of the other.

For an arbitrary function $F(\Phi)$, we find that

$$\frac{d}{dt}F(\Phi) = i[H(\varphi), F(\Phi)] \tag{56}$$

so that the dissipative coupling does not influence time evolution of the SQUID flux variables. It will, however, in general, influence functions of its conjugate variable Q. In fact, we have

$$< \dot{Q} > = < i[H(\varphi), Q] > + < \delta\dot{Q} > \tag{57}$$

where

$$< \delta\dot{Q} > = \int Tr(Qf(\Phi - \phi_{ext} - a)\rho(\varphi, \pi)f(\Phi - \phi_{ext} - a)d\varphi\, d\pi\, da$$

$$-\alpha < Q >= 0 \tag{58}$$

because $Qf - fQ \equiv f'$ and $f'f$ is odd, but $< \dot{Q}^2 >$ can be already $\neq 0$

4.2. The partially deterministic stochastic process associated to SQUID-model

The total Liouville operator of the squid tank model splits as seen in Section 4.1 into 3 parts

$$\mathcal{L} = \mathcal{L}_q + \mathcal{L}_{cl} + \mathcal{L}_{int} . \tag{59}$$

The parts $\mathcal{L}_q$ and $\mathcal{L}_{cl}$ give us deterministic motion of pure states of the quantum and of the classical system. The time evolution is subject to the following coupled system:

$$
\begin{aligned}
i\frac{d\Psi}{dt} &= H(\varphi)\Psi \\
\frac{d\Psi}{dt} &= \pi \\
\frac{d\pi}{dt} &= -\frac{\pi}{R_t C_t} - \frac{\varphi}{L_t C_t} .
\end{aligned}
\tag{60}
$$

They give us a vector field X acting on the product of pure states of the quantum and of the classical system. We write now $\mathcal{L}_{int}$ as acting on observables

$$\int Tr(\rho(\varphi, \pi)(\mathcal{L}_{int}A)(\varphi, \pi)) = \int Tr((\mathcal{L}_{int}\rho)(\varphi, \pi)A(\varphi, \pi)) . \tag{61}$$

After a change of variables and using the cyclicity of the trace this term is then given by

$$\int Tr(\rho(\varphi, \pi)\left[\int f(\Phi - \phi_{ext} - a)A(\varphi, \pi + ka)f(\Phi - \phi_{ext} - a) - \alpha A(\varphi, \pi)\right] . \tag{62}$$

Thus

$$
(\mathcal{L}_{int}A)(\varphi, \pi) = \int da f(\Phi - \phi_{ext} - a)A(\varphi, \pi + ka)f(\Phi - \phi_{ext} - a) \\
-\alpha A(\varphi, \pi) \tag{63}
$$

In order to construct a PD-process we compute now time evolution of functions

$$F_A(\Psi; \varphi, \pi) = (\Psi, A(\varphi, \pi)\Psi) \tag{64}$$

we get

$$
\dot{F}_A(\Psi, \varphi, \pi) = (\Psi, (\mathcal{L}_{int}A)(\varphi, \pi)\Psi) = \tag{65}
$$

$$
= \int da(f\Psi, A(\varphi, \pi + ka)f\Psi) - \alpha F_A(\Psi; \varphi, \pi) =
$$

$$
= \int da\|f\Psi\|^2 F_A\left(\frac{f\Psi}{\|f\Psi\|}, \varphi, \pi + ka\right) - \alpha F_A(\Psi, \varphi, \pi)
$$

$$
= \int da\|f\Psi\|^2\delta(\Psi' - \frac{f\Psi}{\|f\Psi\|})\delta(\varphi' - \varphi)\delta(\pi' - \pi - ka)F_A(\Psi'; \varphi', \pi') - \alpha F_A(\Psi, \varphi, \pi) .
$$

We write the integral kernel as

$$\tilde{Q}(\Psi, \varphi, \pi | \Psi', \varphi', \pi') = \int da \|f\Psi\|^2 \delta \left(\Psi' - \frac{f\Psi}{\|f\Psi\|} \right)$$
$$\times \qquad \delta(\varphi' - \varphi)\delta(\pi' - \pi - ka) \ . \tag{66}$$

We can now perform the a integration and obtain

$$\tilde{Q}(\Psi, \varphi, \pi | \Psi', \varphi', \pi') = \frac{1}{k} \|\tilde{f}\Psi\|^2 \delta \left(\Psi' - \frac{\tilde{f}\Psi}{\|\tilde{f}\Psi\|} \right) \delta(\varphi' - \varphi) \tag{67}$$

where $\tilde{f} = f(\Phi - \phi_{ext} - \frac{\pi' - \pi}{k})$.
We compute next the rate function

$$\lambda_{\varphi,\pi}(\Psi) = \int d\Psi' \int d\varphi' \int d\pi' \, \tilde{Q}(\Psi, \varphi, \pi | \Psi', \varphi', \pi') =$$
$$= \frac{1}{k} \int d\pi' \, \|\tilde{f}\Psi\|^2 = \alpha \ . \tag{68}$$

Then introducing $\tilde{Q}$ by

$$Q(\Psi; \varphi, \pi | \Psi'; \varphi', \pi') = \frac{\tilde{Q}}{\alpha} \tag{69}$$

we obtain for $\dot{F}_A$

$$\dot{F}_A(\Psi, \varphi, \pi) = \alpha \int Q(\Psi, \varphi, \pi | \Psi', \varphi', \pi') F_A(\Psi', \varphi', \pi')$$
$$- \alpha \, F_A(\Psi, \varphi, \pi) \ . \tag{70}$$

This time evolution equation is obviously of the type discussed by Davis [23]. As a last remark we note that the partially deterministic time evolution can be described in the following way:
The system starts in the pure state $(\Psi_0, \varphi_0, \pi_0)$ and evolves deterministically – the classical system according to

$$\ddot{\varphi} + \frac{\dot{\varphi}}{R_t C_t} + \frac{\varphi}{L_t C_t} = 0 \tag{71}$$

and the quantum system according to

$$i\dot{\Psi} = H(\varphi)\Psi \tag{72}$$

until random time t_1 governed by a Poisson process with constant rate α. At time t_1 quantum state jumps to

$$\frac{f(\Phi - \phi_{ext}(\varphi) - \frac{\pi' - \pi}{k})\Psi}{\|f(\Phi - \phi_{ext}(\varphi) - \frac{\pi' - \pi}{k})\Psi\|} \tag{73}$$

and the classical system changes its state in the following way

$$\varphi \longrightarrow \varphi' = \varphi \tag{74}$$
$$\pi \longrightarrow \pi' \tag{75}$$

with probability density

$$\frac{1}{k}\|f(\Phi - \phi_{ext} - \frac{\pi' - \pi}{k})\Psi\|^2 \ . \tag{76}$$

Notice that $\varphi' = \varphi$ from which it follows that the trajectories of the classical system are continuous. Only velocity $\pi = \dot{\varphi}$ jumps at random jump times.

5. Concluding remarks

The mathematical developments constituting Quantum Mechanics have been outstandingly successful in describing and computing (although we would not say explaining) not only those phenomena for which it was invented but also numerous others making many wonderful advances in technology possible. On the other way it is fair to say that the conceptual basis of Quantum Mechanics is still somewhat obscure. The class of models we introduced seems to provide a reliable means of extracting from mathematically consistent models of information transfer from $\sum_q$ to $\sum_c$ well defined predictions for the outcome of any experiment we can envisage - apart of course from the difficulty of solving the mathematical equations, which can be intricate and sophisticated. Physics is the study of reproducible phenomena and a statistical theory of the quantum world is all that theoretical physics whould seek. But as a statistical theory Quantum Mechanics is still a deterministic theory. On the other hand recent advances in study of chaos and algorithmic randomness suggest that near future can bring essentially new elements to our understanding of randomness - both in the realm of foundations of science and in the Nature itself. Any progress in this area may influence our current quantum paradigm.

References

1. J. S. Bell, Against "Measurement " Physics World, August 1990, 33-40.
2. Ph. Blanchard, A. Jadczyk, On the interaction between classical and quantum systems, Physics Letters A 175 (1993) 157-164.
3. Ph. Blanchard, A. Jadczyk, Classical and Quantum Intertwine, in Proceedings of the Symposium of Foundations of Modern Physics, Cologne, June 1993, Ed. P. Mittelstaedt, World Scientific (1993).
4. Ph. Blanchard, A. Jadczyk, From Quantum Probabilities to Classical Facts, in Advances in Dynamical Systems and Quantum Physics, Capri, May 1993, Ed. R. Figari, World Scientific (1994).
5. R. Haag, Irreversibility introduced on a fundamental level, Comm. Math. Phys. 123 (1990) 245-251.
6. J. L. Sauvageot, J. P. Thouvenot, Une nouvelle définition de l'entropie dynamique des systémes non commutatifs, Comm. Math. Phys. 145 411-423 (1992).
7. P. Busch, P. J. Lahti, P. Mittelstaedt, The Quantum Theory of Measurement, Springer (1991).
8. M. Namiki, S. Pascazio, Quantum Theory of Measurement Based on the many-Hilbert-Space approach, Physics Reports 232 No. 6 (1993) 301-411.
9. N. G. van Kampen, Macroscopic Systems in Quantum Mechanics, Physica A 194 (1993) 542-550.
10. J. S. Bell, Are these quantum jumps? in Schrödinger, Centenary of a Polymath", Cambridge University Press (1987).
11. Ph. Blanchard, A. Jadczyk, Strongly coupled quantum and classical systems and Zeno's effect, Physics Letters A 183 (1993) 272-276.

12. A. Connes, H. Narnhofer, W. Thirring, Dynamical entropy of C^*-algebras and von Neumann algebras, Comm. Math. Phys. 112, 691-719 (1987).
13. Ph. Blanchard, A. Jadczyk, Coupled quantum and classical systems, measurement process, Zeno effect and all that, in preparation.
14. Ph. Blanchard, A. Jadczyk, Nonlinear effects in coupling between classical and quantum systems: SQUID coupled to a classical damped oscillator, to appear.
15. A. Jadczyk, Topics in Quantum Dynamics, to appear in Proc. First Caribbean Spring School of Math. and Theor. Phys., Saint Francois, Guadeloupe June 1993, World Scientific, to appear
16. Y. Aharonov, J. Anadan, L. Vaidman, Meaning of the wave function, Phys. Rev. A47 (1993) 4616-4626.
17. Y. Aharonov, L. Vaidman, Measurement of the Schrödinger wave of a simple particle, Phys. Letters A178 (1993) 38-42.
18. T. P. Spiller, T. D. Clark, R. J. Prance, H. Prance, The adiabatic monitoring of quantum objects, Phys. Letters A170 (1992) 273-279.
19. T. P. Spiller, T. D. Clark, R. J. Prance, A. Widom, Quantum Phenomena in Circuits at low temperature, Progress in Low Temperature Physics, Vol. XIII (1992) 221-265.
20. J. F. Ralph, T. P. Spiller, T. D. Clark, R. J. Prance, H. Prance, Chaos in a coupled quantum-classical system, Phys. Letters A180 (1993) 56-60.
21. J. F. Ralph, T. P. Spiller, T. D. Clark, H. Prance, R. J. Prance, A. J. Clippingdale, Non-linear behaviour of teh rf-SQUID magnometer, Physica D63 (1993) 191-201.
22. Dekker,H.:On the phase space quantization of the linearly damped harmonic oscillator, Physica 95 A (1979) 311 – 323
23. Davis,M.H.A., Markov models and optimization, Chapman and Hall, London 1991

CHAOTIC DYNAMICS IN A PERIODICALLY DRIVEN ANHARMONIC OSCILLATOR

YU.L. BOLOTIN
*Kharkov Institute of Physics and Technology, the Ukrainian Academy of Sciences, Akademicheskaja St.1, 310108 Kharkov, Ukraine

V.YU. GONCHAR*
Scientific and Technological Center of Electrophysics, the Ukrainian Academy of Sciences, p/o box 9410 Valter St.3, 310108 Kharkov, Ukraine

and

M.YA. GRANOVSKY
GNPO "Metrology", Mironositskaja St.42, 310078 Kharkov, Ukraine

Abstract. Analytical and numerical studies of the classical dynamics of a periodically driven oscillators show the existence of the transition regularity- chaos-regularity. This means that not only at high energy but also at low energy conserve isolated nonlinear resonances in the vicinity of which the motion remains regular.

The one–dimensional nonlinear Hamiltonian system with time–dependent perturbation is the simplest dynamical system admitting chaotic behavior. An external field induces a dense set of resonance zones in the phase space of dynamical system. For weak external fields the principal resonance zones remain isolated: the Kolmogorov–Arnold–Moser (KAM) surface prevents the particle trajectories from wandering from one resonance to another. As the perturbation increases KAM invariants between neighboring zones begin to be destroyed until a pathway is finally opened. When this happens one can speak [1] about the onset of a global stochasticity. This qualitative picture provides a means for estimating the size of the perturbation required to make the transition from regular to stochastic motion. Roughly speaking, this occurs when the neighboring resonances overlap. A diffusive energy transfer is the hallmark of deterministic chaos in its classical version. Numerous theoretical and experimental investigations of this process have been made for different physical systems such as: the Rygberg atom in a microwave field [2], a particle in quartic [3] and periodic [4] potentials with a time–dependent perturbation, etc. The general peculiarity of these systems is that the density of resonances increases as a function of energy. If the particle has an initial energy, which lies between k and k+1 resonances, then the application of an external field larger than the threshold, defined, for example, by the resonance overlap criterion [5] will cause the particle to diffuse in energy (action). Moreover, since the resonance overlapping increases with k, once the external field exceeds the threshold for stochastic diffusion for a particle with energy E_k, then the confining KAM surfaces will also be destroyed for larger energy. Since the stochastic region in the phase space is bounded below and unbounded above, the particle will tend to diffuse to larger energies (actions). The goal of this paper is to consider a radically different situation in which the density

31

Z. Haba et al. (eds.), Stochasticity and Quantum Chaos, 31–37.
© 1995 *Kluwer Academic Publishers.*

of resonances decreases as a function of energy. In this case, for a given external field the accessible phase space will be bounded both below and above. Therefore in the transition to global stochasticity, diffusion will also be bounded in the finite part of the phase space. The situation with the increasing distances between resonances may be realized in the potential $U(x) = Ax^n$ $(n = 2l, \ l > 1)$.

Let us consider the dynamics of a periodically driven anharmonic oscillator generated by the Hamiltonian

$$H(p, x, t) = H_0(p, x) + F \, x \cos \Omega \, t \qquad (1)$$

where the unperturbed Hamiltonian is given by

$$H_0(p, x) = \frac{p^2}{2m} + Ax^n = E \qquad (n = 2l, \ l > 1) \ .$$

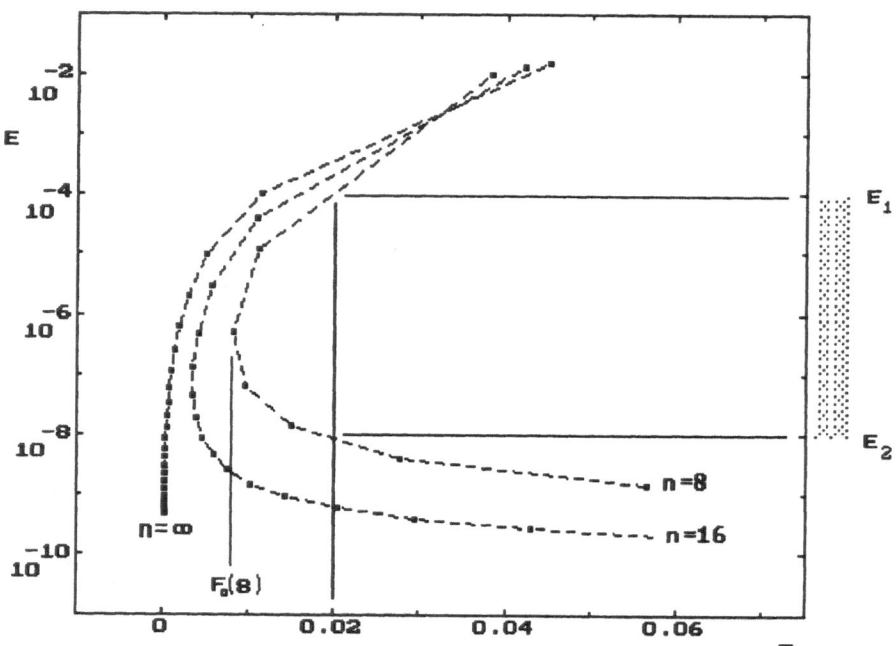

Fig. 1. Separatrices $E(n, F)$, determinated by equations (9)–(16), divide the phase space of the system (1) into regular and stochastic ranges. Numerical results for the shaded energy interval $E_1 - E_2$ will be shown on Figs.2 and 3.

Before considering the effects of a perturbation it is convenient to make a canonical transformation to action–angle variables (I, Θ)

$$I = \frac{G(n)}{\pi} \frac{n}{n+2} E^{\frac{n+2}{2n}}$$

$$\Theta = \begin{cases} \pi \left[1 - sign(x - x_0)\right] + \bar{\Theta} & p > 0 \\ \\ \pi \left(1 - x_0\right) - \bar{\Theta} & p < 0 \end{cases} \qquad (2)$$

Here

$$G(n) = \frac{2\sqrt{2\pi m}}{A^{1/n}} \frac{\Gamma(1+1/n)}{\Gamma(1/2+1/n)}$$

$$\bar{\Theta} = \sqrt{\frac{m}{2}} \frac{2n}{n+2} \frac{E^{1/2}}{I} \left(\frac{E}{A}\right)^{1/n} \int_{x_0(A/E)^{1/n}}^{x(A/E)^{1/n}} \frac{dy}{\sqrt{1-y^n}} .$$

(3)

We will now obtain estimates for the amplitude E_k^{cr}, at which the k and $k+1$ resonance zones overlap. If we expand the perturbation in a Fourier series in Θ, the perturbed Hamiltonian can be written as

$$H(I,\Theta,t) + H_0(I) + \sum_{k=-\infty}^{k=\infty} x_k(I) \cos(k\Theta - \Omega\, t) ,$$

(4)

where the Fourier amplitudes x_k are defined by the integrals

$$x_k(I) = \frac{1}{2\pi} \int_0^{2\pi} d\Theta\, e^{ik\Theta}\, x(I,\Theta)$$

(5)

and the unperturbed Hamiltonian

$$H_0(I) = \left(\frac{n+2}{n} \frac{\pi}{G(n)} I\right)^{\frac{2n}{n+2}} .$$

(6)

As $n \to \infty$ the unperturbed Hamiltonian turns into the well-known Hamiltonian of a square well with width equal to 2,

$$\lim_{n\to\infty} H_0(l) = \frac{\pi^2\, I^2}{8m} .$$

(7)

For small F, the primary resonances are located at the values $I \equiv I_k$ $(E \equiv E_k)$ which satisfy the equation

$$k\omega(I) = \Omega, \qquad \omega(I) \equiv \frac{\partial H_0}{\partial I} .$$

(8)

The resonant actions and energies are therefore determined by the relations

$$I_k = \frac{2n}{n+2} \left(\frac{G(n)}{2\pi}\right)^{\frac{2n}{n-2}} \left(\frac{\Omega}{k}\right)^{\frac{n+2}{n-2}}$$

$$E_k = \left(\frac{G(n)}{2\pi}\right)^{\frac{2n}{n-2}} \left(\frac{\Omega}{k}\right)^{\frac{2n}{n-2}} .$$

(9)

The classical treatment, which is based on the resonance overlap criterion [5] for the onset of global stochasticity, leads to the following condition for the determination of the critical amplitude F_k^{cr},

$$(W_k - W_{k-1})/2 = I_k - I_{k+1}$$

(10)

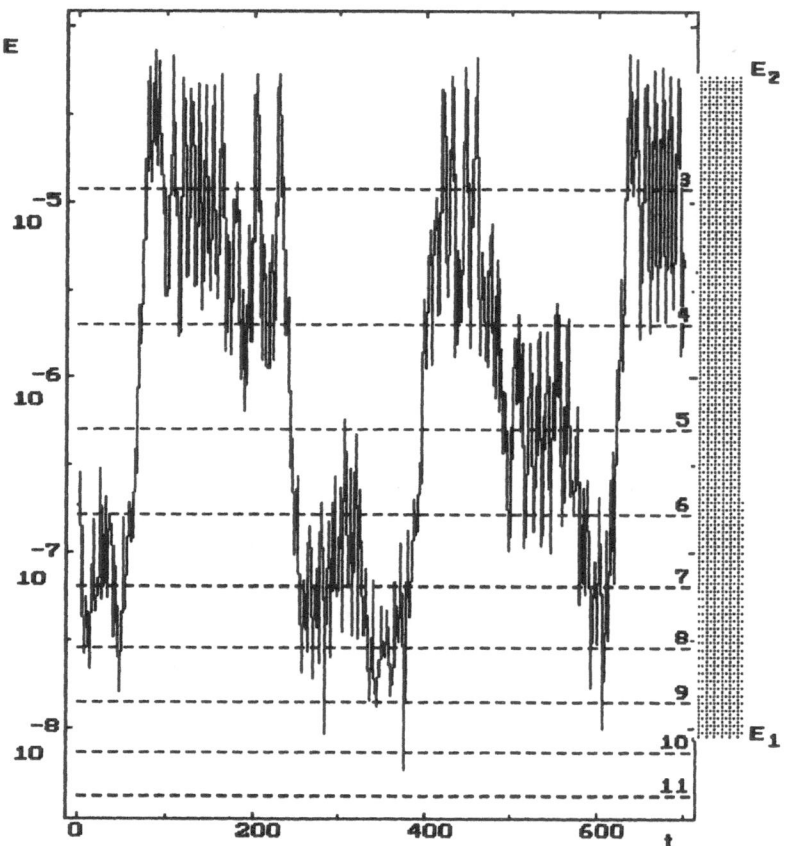

Fig. 2. Long-time behavior $E(t)$ for a single trajectory $x(t)(x_0 = -0.244, p_0 = -0.049, t$ in units $T = 2\pi/\Omega$, $n = 8$, $F = 0.02$, $\Omega = 0.9$). The range of the stochastic diffusion corresponding to analytical evaluations (see Fig. 1) is shaded. The right scale consists of resonance numbers.

where the W_k standing for the width of the k–th nonlinear resonance, are determined by the corresponding Fourier amplitudes x_k,

$$W_k = 4 \left(\frac{F\, x_k(I)}{\omega'(I)} \right)^{1/2} \Bigg|_{I=I_k} . \qquad (11)$$

The Fourier components can be determined by evaluating the integrals (5)

$$x_{2s+1} = \frac{4}{(2s+l)\,\pi} \; E^{1/n} \int_0^1 \cos\left[(2s+1)\; \frac{\pi}{2}\; \frac{f^{(n)}(t)}{f^{(n)}(I)}\right] \, dt \tag{12}$$

$$x_{2s} = 0 \;,$$

where

$$f^{(n)}(t) \equiv \int_0^t \frac{dt'}{\sqrt{1-t'^n}} \; = \; \sum_{k=0}^\infty \frac{(2k+1)!}{(2^k\,k!)^2} \; \frac{t^{kn+1}}{kn+1} \;. \tag{13}$$

The widths of the even resonances are zero to first-order in the constant F. However, they appear in the next-to-leading order. For finding these widths, the ordinary procedure [7] of canonical transformation may be used. Using expressions (9) and (11), the critical field strength can be written as

$$F_k^{cr} = \frac{1}{4} \left[\frac{\Phi\left(\Omega, n\right)\left[k^{\frac{2+n}{2-n}} - (k+1)^{\frac{2+n}{2-n}}\right]}{\left(\frac{x_k}{\omega'\left(I_k\right)}\right)^{1/2} + \left(\frac{x_{k+1}}{\omega'\left(I_{k+1}\right)}\right)^{1/2}} \right]^2, \tag{14}$$

where

$$\Phi(\Omega, n) = \frac{2n}{n+2} \left(\frac{G(n)}{2\pi}\right)^{\frac{2n}{n-1}} \Omega^{\frac{n+2}{n-2}} \;. \tag{15}$$

To the first-order of the perturbation theory in the constant F (zero widths of the even resonances), we find

$$F_k^{cr} = \frac{\omega'\left(I_k\right)}{4x_k} \; \Phi^2(\Omega, n) \; \left[k^{\frac{2+n}{2-n}} - (k+1)^{\frac{2+n}{2-n}}\right]^2. \tag{16}$$

In the limit $n \rightarrow \infty$ (for the square well), we have [8]

$$F_k^{cr} = \frac{\Omega^2}{8(k+1)^2} \;. \tag{17}$$

Figure 1 shows the "phase diagram" that allows us to determine the energy intervals of regular and chaotic classical motion at a fixed level of the periodic perturbation. We shall dwell on its main peculiarities. For the square well ($n = \infty$) and any F value one can indicate the energy value above and below which the motion will be regular and chaotic respectively. When $n < \infty$ the value of the external perturbation $F_0(n)$ can be indicated such that for $F < F_0(n)$ the motion will be regular at all energy values. For any $F > F_0(n)$ the finite energy interval $[E_1(n), E_2(n)]$ can be found, within which the motion will be chaotic as a result of the overlap of nonlinear resonances. The particle with initial energy E_0 of the interval $[E_1(n) \; E_2(n)]$ will randomly wander in this energy interval. A typical diffusion dependence of energy

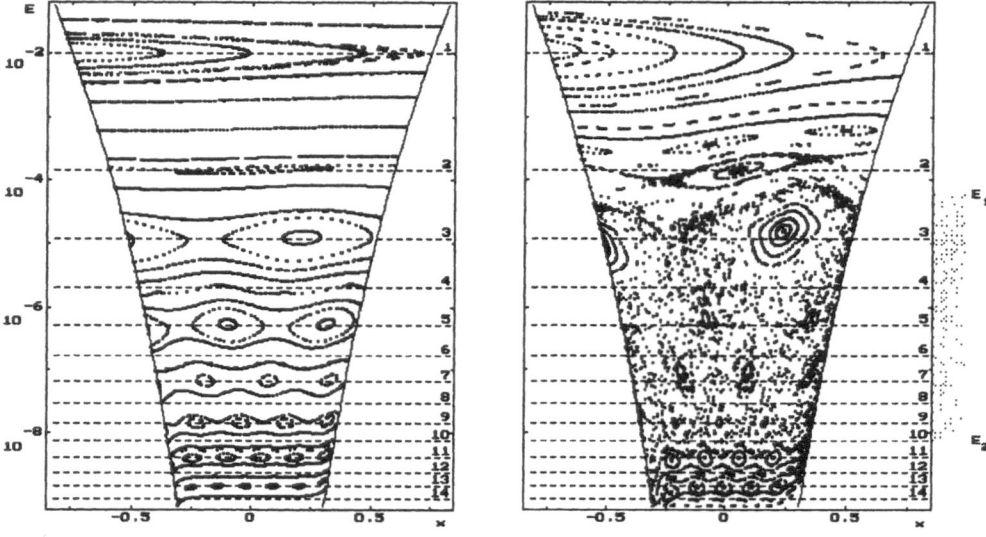

Fig. 3.   Stroboscopic plots for the system (1): (a) $F = 0.007 < F_0(8)$, (b) $F = 0.02 > F_0(8)$; in the shaded energy range (b) all KAM invariants are destroyed and stochastic diffusion on the resonance interval 3-9 takes place, as well as in Fig. 2. Solid curves – potential function x^8, dashed lines – resonances.
Note: Use of the (E, x) variables instead of the traditional (p, x) variables allows us to see the KAM invariants clearly.

versus time is shown in Fig 2. The structure of the phase space for the case $F < F_0(8)$ is shown in Fig. 3a. The difference in the phase space structure of regular and chaotic energy intervals for $F > F_0(8)$ is illustrated in Fig. 3b.

We have considered the dynamics of a system representing a comparatively simple potential well with an external periodic perturbation. We have been interested in the regular–to–chaotic transition as the external perturbation increases, and, particularly in the diffusive energy transfer as one of most interesting manifestations of this transition. As opposed to the potentials studied previously (Coulomb potential, symmetrical double well, cosine–type periodic potential) in the considered system the resonance density increases as the energy decreases. As a result, this structure of the phase space for all n values causes the boundedness of the global stochasticity region from above. However, it appears quite unexpected that in the case of finite n at a fixed external field the energy interval, corresponding to chaotic motion, is bounded from below as well. This means that at low energy conserve isolated nonlinear resonances in the vicinity of which the motion remains regular. It should be note in conclusion the analogy between present problem and autonomous two–dimensional Hamiltonian system. As demonstrated recently [9] in the case of potentials having a localized instability region (in particular, localized region of negative Gaussian curvature) the regular–chaos–regularity transition takes place with energy increase. Fig. 2 [9]. shows "phase diagram" for the Hamiltonian

$$H\left(p_x, p_y, x, y\right) = \left(\frac{P_x^2}{2} + \frac{P_y^2}{2}\right) + \left(\frac{x^2}{2} + \frac{y^2}{2}\right) + b \cdot \left(x^2 y - \frac{y^3}{3}\right) + c \cdot \left(x^2 + y^2\right)^2 .$$

$$(18)$$

Attention is drawn to the fact that the "phase diagram" of Fig. 1 of the one - dimensional potential well with periodic perturbation and that of Fig. 2 [9] of the autonomous two–dimensional Hamiltonian system are very similar. In the first case the amplitude of the periodic perturbation plays the part of the nonlinearity parameter $W = b^2/4c$. We have demonstrated [9,10] for Hamiltonian (18) the existence of a rigid correlation between the character of the classical motion and the statistical properties of the quantum energy spectra and eigenfunctions. It appears natural to investigate similar quantum manifestations of classical stochasticity in both the quasienergy spectrum and the properties of the quasienergy wave functions of the Hamiltonian considered in the present paper. The results of such investigation will be the subject of subsequent publications. This work was supported by State Scientific and Technological Affair Committee of Ukraine.

References

1. G.M. Zaslavskij, Stokhastichnost' dinamicheskikh system (Nauka, Moskow, 1984).
2. G. Casati, B. Chirikov, D. Shepelansky and I. Guarneri, Phys.Rep. **154** (1987)79.
3. L.E. Reichl and W.H. Zheng, Phys. Rev. **A29** (1984)2186.
4. W.A. Lin and L.E. Reichl, Phys. Rev. **A31** (1985)1136.
5. B.V. Chirikov, Atomnaya energya 6 (1959)630.
6. H. Breuer, K. Dietz, M. Holthaus, Physica **D46** (1990)317.
7. A. Lichtenbegr and M. Liberman, Regular and stochastic motion (Springer Verlag N.Y. 1983).
8. W.A. Lin and L.E. Reichl, Physica **D19** (1986)145.
9. Y.L. Bolotin et al., Phys. Lett. **135** (1989)29.
10. Y.L. Bolotin et al., Phys. Lett. **144** (1990)459.

COHERENT AND INCOHERENT DYNAMICS IN A PERIODICALLY DRIVEN BISTABLE SYSTEM

T. DITTRICH, F. GROSSMANN, P. HÄNGGI, B. OELSCHLÄGEL and
R. UTERMANN
Institut für Physik, Universität Augsburg,
Memminger Straße 6, D-86135 Augsburg, Germany

Abstract. We study the conservative as well as the dissipative quantal dynamics in a harmonically driven, quartic double-well potential. Our main tool is a numerical analysis of time evolution and spectrum, based on the Floquet formalism. In the deep quantal regime, we find coherent modifications of tunneling, including its complete suppression. In the semiclassical regime of the conservative system, the dynamics is dominated by the competition of tunneling between symmetry-related regular regions and chaotic diffusion along the separatrix. We demonstrate that there is a strong correlation between each tunnel splitting and the overlap of the associated doublet states with the chaotic layer. In the dissipative case, remnants of coherent behavior occur as transients, such as the tunneling between distinct, coexisting limit cycles. In particular, the coherent suppression of tunneling is stabilized by weak incoherence. The quantal stationary states are characterized by an anisotropical broadening due to quantum noise, as compared to the corresponding classical attractors.

Key words: quantum chaos, bistable systems, tunneling, Floquet theory, semiclassical theory, dissipative quantum chaos

1. Introduction

A good way to learn about the interplay of coherent and incoherent dynamics, of stochasticity and deterministic chaos is to study specific systems where they occur simultaneously. Bistable systems form an important paradigm: They contain the essence of a wealth of nonlinear phenomena, from the microscopic to the macroscopic realm [1 – 5]. With a periodic driving added, their classical repertoire of behavior ranges from limit cycles to several coexisting strange attractors [6 – 9]. On the quantum-mechanical level, bistable systems provide the standard example of coherent tunneling [10].

In the present paper we investigate the nature of the transition from the simple coherent quantal dynamics to the intricate complexity of the macroscopic behavior, focussing on a few selected landmarks. This transition has two basic aspects: One of them is the sheer increase in system size. In the formal description, it is reflected in an increase of characteristic actions, compared to $\hbar$, and thus corresponds to the short-wavelength or semiclassical limit. The other aspect is the growing importance of ambient and internal degrees of freedom, weakly coupled to the system in focus, and usually modelled collectively as a reservoir.

The transition to short characteristic wavelengths lets classical phase-space structures emerge more and more clearly in the quantal dynamics. Specifically, in the case of bistable systems, we shall ask how the onset of chaos on the classical level

39

Z. Haba et al. (eds.), Stochasticity and Quantum Chaos, 39–55.
© 1995 *Kluwer Academic Publishers.*

becomes manifest in the tunneling.

Incoherent processes induced by the ambient degrees of freedom, in turn, tend to smooth out the fine interference patterns in phase space and time which encode classical behavior on the quantal level, and thus render the semiclassical limit less singular. We study the stationary states approached by the dissipative quantum system—they correspond to the attractors of the classical flow—and the decay of coherence in the transient behavior preceeding them.

We shall introduce our working model, the harmonically driven quartic double well, and its symmetries in Section 2. Section 3 is devoted to the modifications of tunneling, due to the driving, in the deep quantal regime. In Section 4, we discuss driven tunneling in the semiclassical regime where it begins to exhibit the influence of classical chaos. The consequences of incoherent processes, of damping and noise, are addressed in Section 5. Section 6 summarizes our survey of coherent and incoherent behavior in bistable systems.

The present work forms a synopsis of results partially published elsewhere [11 – 19].

2. The model and its symmetries

The harmonically driven quartic double well is described by the Hamiltonian

$$H_{\mathrm{DW}}(x,p;t) = H_0 + H_1, \quad H_0(x,p) = \frac{p^2}{2} - \frac{1}{4}x^2 + \frac{1}{64D}x^4, \ H_1(x;t) = x\, S \cos\omega t. \quad (1)$$

With the dimensionless variables used, the only parameter controlling the unperturbed Hamiltonian $H_0(x,p)$ is the barrier height D. It can also be interpreted as the (approximate) number of doublets with energies below the top of the barrier. Accordingly, the classical limit amounts to letting $D \to \infty$. The driving is characterized by its amplitude S and frequency ω.

The symmetry of the Hamiltonian under discrete time translations, $t \to t + 2\pi/\omega$, enables to use the Floquet formalism [20 – 23], which generalizes most of the conceptual tools of spectral analysis to the present context. Its basic ingredient is the Floquet operator, i.e., the unitary propagator that generates the time evolution over one period of the driving force,

$$U = \mathsf{T} \exp\left(-\frac{\mathrm{i}}{\hbar} \int_0^{2\pi/\omega} \mathrm{d}t\, H_{\mathrm{DW}}(t)\right), \quad (2)$$

where T effects time ordering. Its eigenvectors and eigenphases, referred to as *Floquet states* and *quasienergies*, respectively, can be written in the form

$$|\psi_\alpha(t)\rangle = \mathrm{e}^{-\mathrm{i}\epsilon_\alpha t}|\phi_\alpha(t)\rangle, \quad \text{with } |\phi_\alpha(t + 2\pi/\omega)\rangle = |\phi_\alpha(t)\rangle. \quad (3)$$

From a Fourier expansion of the $|\phi_\alpha(t)\rangle$,

$$|\phi_\alpha(t)\rangle = \sum_k |\phi_{\alpha,k}\rangle \mathrm{e}^{-\mathrm{i}k\omega t}, \quad |\phi_{\alpha,k}\rangle = \frac{\omega}{2\pi}\int_0^{2\pi/\omega}\mathrm{d}t\,|\phi_\alpha(t)\rangle \mathrm{e}^{\mathrm{i}k\omega t}, \quad (4)$$

it is obvious that the quasienergies are organized in classes, $\epsilon_{\alpha,k} = \epsilon_\alpha + k\omega$, $k = 0, \pm 1, \pm 2, \ldots$, where each member corresponds to a physically equivalent solution. Therefore, all spectral information is contained in a single "Brillouin zone", $-\omega/2 \leq \epsilon < \omega/2$.

Besides invariance under time translation and time reversal, the unperturbed system possesses the spatial reflection symmetry $x \rightarrow -x$, $p \rightarrow p$, $t \rightarrow t$. For a general driving, this symmetry is destroyed. For the specific time dependence of a harmonic driving, however, the symmetry $f(t + \pi/\omega) = -f(t)$ restores a similar situation as in the unperturbed case: The system is now invariant against the operation [11 – 13, 24]

$$\mathsf{P} : p \rightarrow -p, \quad x \rightarrow -x, \quad t \rightarrow t + \frac{\pi}{\omega}, \tag{5}$$

which may be regarded as a *generalized parity* in the extended phase space spanned by x, p, and phase, i.e., time $t \bmod(2\pi/\omega)$. As in the unperturbed case, this enables to separate the eigenstates into an even and an odd subset.

3. Driven tunneling and localization

To give an impression of driven tunneling in the deep quantal regime, we study in the following how a state, prepared as a coherent state centered in the left well, evolves in time under the external force. Since this state is approximately given by a superposition of the two lowest unperturbed eigenstates, $|\Phi(0)\rangle \approx (|\Psi_1\rangle + |\Psi_2\rangle)/\sqrt{2}$, its time evolution is dominated by the Floquet-state doublet originating from $|\Psi_1\rangle$ and $|\Psi_2\rangle$, and the splitting $\epsilon_2 - \epsilon_1$ of its quasienergies.

There are two regimes in the (ω, S)-plane where tunneling is not qualitatively altered by the external force: Both in the limits of slow (adiabatic) and of fast driving, the separation of the time scales of the inherent dynamics and of the external force effectively uncouples these two processes and results in a mere renormalization of the tunnel splitting Δ. Specifically, as both an analytical treatment and numerical experiments show, the driving always reduces the effective barrier height and thus increases the tunneling rate in these two limits [11].

Qualitative changes in the tunneling behavior are expected as soon as the driving frequency becomes comparable to the internal frequencies of the double well, in particular, to the tunnel splitting and to the so-called resonances $E_3 - E_2$, $E_4 - E_1$, $E_5 - E_2$, ... By spectral decomposition, the temporal complexity in this regime is immediately related to the "landscape" of quasienergy planes $\epsilon_{\alpha,k}(\omega, S)$ in parameter space. Features of particular significance are close encounters of quasienergies: Two quasienergies cross one another without disturbance if they belong to different parity classes, otherwise they form an avoided crossing.

A quantity well suited to study the relationship between dynamics and quasienergy spectrum is the *probability to return* [25, 26],

$$P^\Phi(t_n) = |\langle \Phi(0) | \Phi(t_n)\rangle|^2 = |\langle \Phi(0) | U^n | \Phi(0)\rangle|^2, \quad t_n = \frac{2\pi n}{\omega}, \quad n = 0, \pm 1, \pm 2, \ldots, \tag{6}$$

defined with respect to some initial state $|\Phi(0)\rangle$, and with time restricted to the instances of zero phase of the driving. By expanding $P^\Phi(t_n)$ in the basis provided

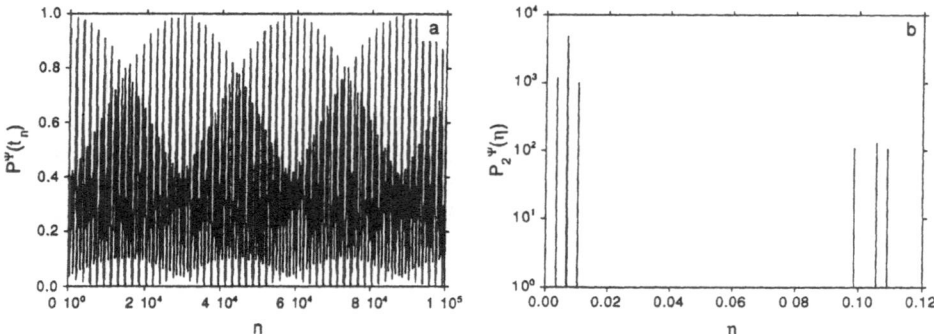

Fig. 1.   Driven tunneling at the fundamental resonance $\omega = E_3 - E_2$. (a) Time evolution of $P^{\Phi}(t_n)$ over the first 2×10^5 time steps; (b) local spectral two-point correlation function $P_2^{\Phi}(\eta)$ obtained from (a). The parameter values are $D = 2$, $S = 2 \times 10^{-3}$, and $\omega = 0.876$.

by the quasienergy eigenstates,

$$P^{\Phi}(t_n) = \xi^{-1} + \sum_{\alpha \neq \beta} \exp[i(\epsilon_\alpha - \epsilon_\beta)t_n] \, |\langle \phi_\alpha(0) \,|\, \Phi(0) \rangle|^2 \, |\langle \phi_\beta(0) \,|\, \Phi(0) \rangle|^2, \quad (7)$$

the rôle of the quasienergies for the time evolution becomes explicit. Here, ξ^{-1} stands for the time-independent diagonal part excluded from the double sum in Eq. (7). It gives the long-time average of $P^{\Phi}(t_n)$. The Fourier transform $P_2^{\Phi}(\eta)$ of the probability to return is the two-point correlation function of the *local* quasienergy spectrum, i.e., the spectrum weighted according to the relative significance of each quasienergy for the specific dynamics starting from $|\Phi(0)\rangle$. Below, we use these concepts to discuss driven tunneling at two parameter points, one of them featuring an avoided quasienergy crossing, the other an exact one [11 – 14].

The "single-photon transition" at $\omega = E_3 - E_2$ is called *fundamental resonance*. At $S = 0$, it is reflected in a crossing between the quasienergies $\epsilon_{2,k}$ and $\epsilon_{3,k-1}$. For $S > 0$, it becomes an avoided crossing, since the corresponding eigenstates have equal parity. Fig. 1a shows the time evolution of $P^{\Phi}(t_n)$ at the fundamental resonance $(D = 2,\ S = 2 \times 10^{-3},\ \omega = 0.876)$. The monochromatic oscillation of $P^{\Phi}(t_n)$ characteristic of unperturbed tunneling has given way to a more complex beat pattern. The Fourier transform of $P^{\Phi}(t_n)$ reveals that these beats are composed mainly of two groups of three frequencies each (Fig. 1b), which can be identified, in turn, as the quasienergy differences $\epsilon_{3,-1} - \epsilon_{2,0}$, $\epsilon_{2,0} - \epsilon_{1,0}$, $\epsilon_{3,-1} - \epsilon_{1,0}$, and $\epsilon_{4,-1} - \epsilon_{3,-1}$, $\epsilon_{4,-1} - \epsilon_{2,0}$, $\epsilon_{4,-1} - \epsilon_{1,0}$, at the avoided crossing.

In contrast, a two-photon transition that bridges the tunnel splitting Δ is "parity forbidden", and thus the quasienergies $\epsilon_{2,k-1}$ and $\epsilon_{1,k+1}$ form an exact crossing. Eq. (7) indicates that a vanishing of the difference $\epsilon_{2,-1} - \epsilon_{1,1}$ will have a remarkable consequence: For a state prepared as an exact superposition of the corresponding two Floquet eigenstates only, $P^{\Phi}(t)$ and all other observables become constants, at least at the discrete times t_n, and thus it is possible that tunneling comes to a standstill! According to an argument going back to von Neumann and Wigner [27, 28], exact crossings should occur along one-dimensional manifolds in the (ω, S)

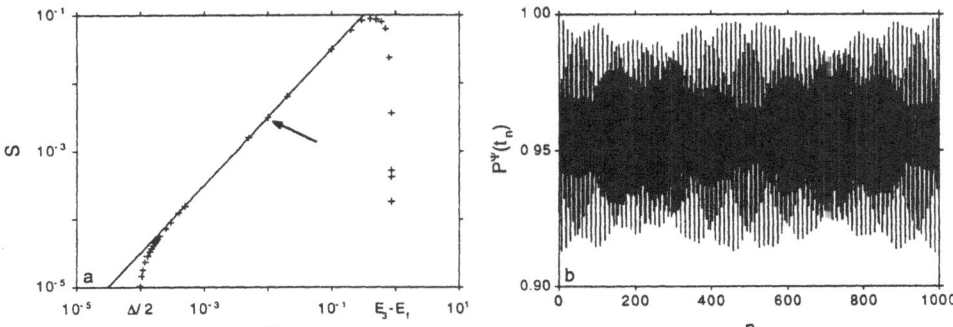

Fig. 2.   Suppression of tunneling at an exact crossing $\epsilon_{2,-1} = \epsilon_{1,1}$. (a) The manifold $S_{\text{loc},0}(\omega)$ in the (ω, S) plane where this crossing occurs (data obtained by diagonalization of the full Floquet operator for the driven double well are indicated by crosses, the full line has been derived from a two-state approximation [15, 29], the arrow indicates the parameter point to which part (b) refers); (b) time evolution of $P^{\Phi}(t_n)$ over the first 10^3 time steps.

plane. In the present case, there is one such manifolds $S_{\text{loc},k}(\omega)$ for each condition $\epsilon_{2,-k} = \epsilon_{1,k}$. Fig. 2a shows $S_{\text{loc},0}(\omega)$: It is a closed curve, reflection-symmetric with respect to the line $S = 0$, with an approximately linear frequency dependence for $\Delta \lesssim \omega \lesssim E_3 - E_2$. A typical time evolution of $P^{\Phi}(t_n)$ for a parameter point ($D = 2$, $S = 3.171 \times 10^{-3}$, $\omega = 0.01$) on the linear part of that manifold is presented in Fig. 2b. It clearly indicates that tunneling is almost completely suppressed. The remaining oscillations of small amplitude can be ascribed to an admixture of higher-lying quasienergy states to the initial state. An additional time dependence faster than the driving—it would not show up in a stroboscopic plot like Fig. 2b—does indeed exist, but with an amplitude comparable to that appearing in Fig. 2b (not shown, see refs. [11 – 14].

The suppression of tunneling is an elementary quantum-interference effect. In fact, much of it can be understood on basis of a two-state approximation [15, 29]. It is achieved by solving the equations of motion for the expansion coefficients of a localized initial state in the Hilbert space spanned by the unperturbed ground-state doublet $|\Psi_1\rangle$, $|\Psi_2\rangle$. The two-state approximation yields an analytical expression for each $S_{\text{loc},k}(\omega)$.

4. Tunnel splittings and the onset of chaos

In the classical double well, the most significant consequence of the periodic driving is the onset of deterministic chaos [30], see Fig. 3. It develops around the hyperbolic fixed point at the top of the barrier: As the perturbation is switched on, the stable and the unstable manifold intersecting at this fixed point start to fold and form a homoclinic tangle [31], which extends all along the two lobes of the separatrix and forms a narrow layer of chaotic motion. With S increasing further, this chaotic layer grows in width, at the expense of the two regular zones within the wells, so that the deterministic diffusion between the wells becomes a significant contribution to

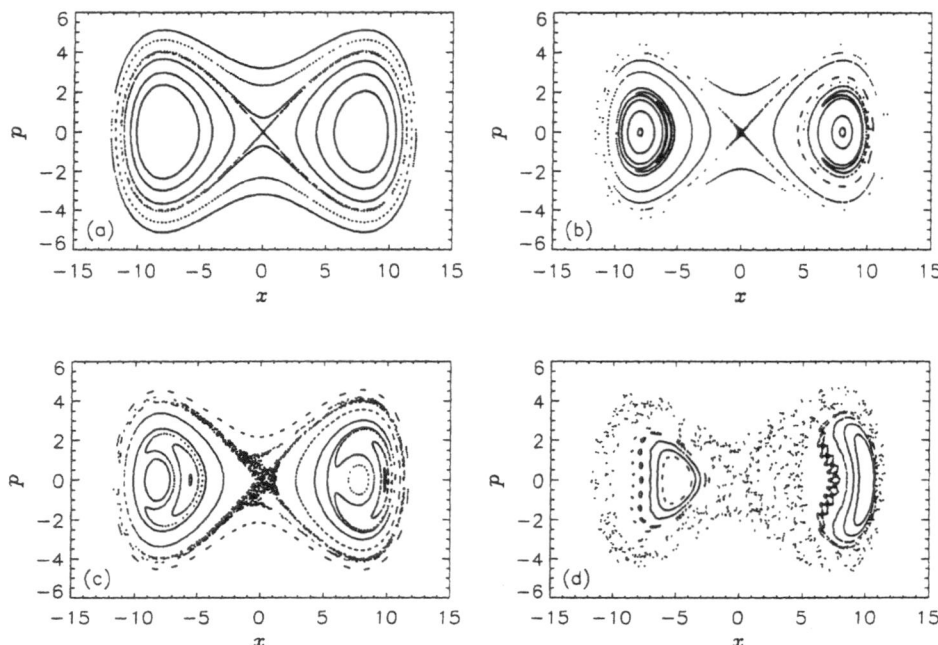

Fig. 3.   Classical phase-space portraits of the periodically driven double well at phase 0 of the driving, for various values of the driving amplitude. The parameter values are $D = 8$, $\omega = 0.95$, and (a) $S = 0$, (b) $S = 10^{-3}$, (c) $S = 10^{-2}$, (d) $S = 0.2$.

the classical phase-space transport. There is another conspicuous modification of the phase-space structure, the growth of the regular zone generated by the first resonance of the driving with the unperturbed oscillation [30], which does not, however, affect the coherent dynamics as substantially as the onset of chaos does.

Quantum mechanically, phase-space transport by chaotic diffusion competes with tunneling [32 – 37]. In the present section we investigate how these two processes influence each other. Applying ideas from Einstein-Brillouin-Keller (EBK) quantization for periodically driven systems [38], as well as from random-matrix theory for mixed (regular and chaotic) systems [39], to the present context, we arrive at the following simple expectation [33 – 35]: Even with the driving, the two isolated regular regions within the wells remain related by a discrete symmetry, the generalized parity P (see Eq. 5). Accordingly, Floquet states residing within these regions should form a more or less regular ladder of tunnel-splitted doublets. For states mainly residing within the chaotic layer, in contrast, random-matrix theory predicts level repulsion. We therefore expect that, as soon as one of the pairs of quantizing tori pertaining to the symmetry-related regular regions resolves in the spreading chaotic layer, the exponentially small splitting of the corresponding doublet widens until it reaches a size of the order of the mean level separation. As a consequence, the coherent tunneling on an extremely long time scale will give way to a more irregular

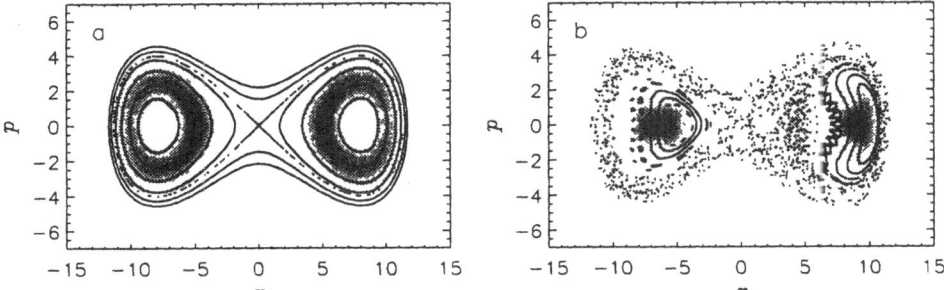

Fig. 4.   Husimi distributions (in gray-scale representation) for the quasienergy state $|\phi_7(0)\rangle$, compared with the corresponding classical phase-space portraits, at (a) $S = 10^{-5}$ and (b) $S = 0.2$.

dynamics on shorter time scales, forming the quantal counterpart of deterministic diffusion along the separatrix.

We emphasize that the breakup of the tunnel doublets in the chaotic layer is not a direct consequence of a local property, the positive Lyapunov exponent. Rather, it depends on the fact that diffusive spreading connects all parts of the chaotic layer, even across the symmetry plane. Furthermore, one should keep in mind that the disintegration of a classical torus, looked at closely, is not an abrupt event but proceeds through an intermediate "leaky" stage with fractal dimension ("cantorus") [40]. Even after the cantorus has disappeared, a distinct repelling structure remains within the chaotic layer ("vague torus") [41].

In order to check our hypothesis numerically, we have to quantify the distinction between "regular" and "chaotic" eigenstates, i.e., states located mainly in regions of a corresponding nature in classical phase space. We base this quantification on a quantum-mechanical probability density in phase space, the Husimi distribution [42, 43]. The overlap of the Husimi representation $Q_\alpha(x, p; t)$ of a Floquet state $|\psi_\alpha(t)\rangle$ with the chaotic layer [18, 19],

$$\bar{\Gamma}_\alpha = \frac{\omega}{2\pi} \int_0^{2\pi/\omega} dt \int_{-\infty}^{\infty} dx \int_{-\infty}^{\infty} dp \, Q_\alpha(x, p; t) \Gamma(x, p; t), \qquad (8)$$

can be used as a measure of "how chaotic that state is". Here, $\Gamma(x, p; t)$ denotes the characteristic function for the chaotic region. It can be determined numerically, e.g., by letting a trajectory started anywhere in this chaotic region "tick" boxes in a coarse-grained phase space of the desired resolution. Since the Husimi distribution forms a normalized probability distribution over phase space, we have $0 \leq \bar{\Gamma}_\alpha \leq 1$. As an illustration of these concepts, we compare, in Fig. 4, the Husimi distribution for the quasienergy state $|\phi_7\rangle$ with the corresponding classical phase space portrait, for (a) $S = 10^{-5}$ and for (b) $S = 0.2$, at phase 0 of the driving.

The simple picture sketched above clearly implies a strong relationship between the tunnel splittings $\Delta_\lambda = |\epsilon_{2\lambda,k} - \epsilon_{2\lambda-1,k}|$ and the overlaps $\bar{\Gamma}_{2\lambda-1} \approx \bar{\Gamma}_{2\lambda}$. In Fig. 5, we compare the S dependences of these two sets of quantities for the seven quasienergy doublets from $|\phi_1\rangle$, $|\phi_2\rangle$ to $|\phi_{13}\rangle$, $|\phi_{14}\rangle$ [18, 19]. Qualitatively, we

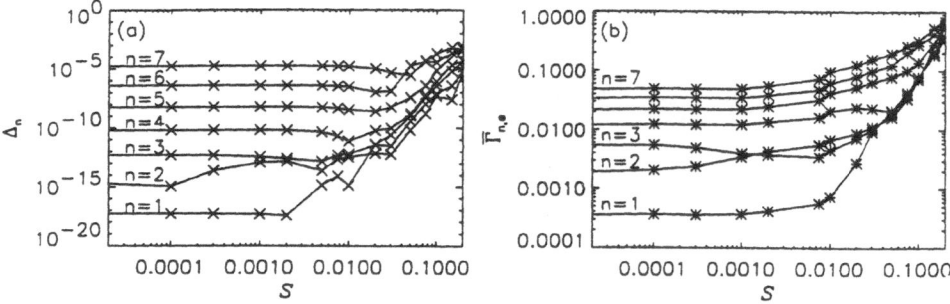

Fig. 5.   Tunnel splittings (a) and overlaps with the chaotic layer (b) for the seven lowest tunnel doublets, as functions of the amplitude of the driving.

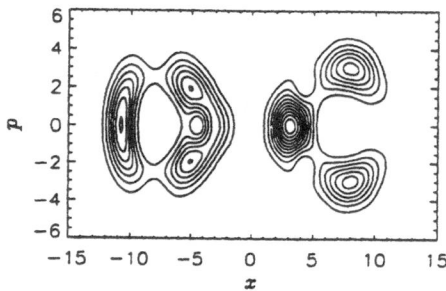

Fig. 6.   Contour plot of the Husimi distribution for the quasienergy state $|\phi_{13}(0)\rangle$, at $S = 0.2$.

observe a striking similarity: There is only a weak S dependence, reflecting the influence of the growing first resonance, for $S \lesssim 10^{-3}$. For $S \gtrsim 10^{-3}$, both the tunnel splittings and the overlaps start to grow exponentially, one by one, starting from the lowest doublet, so that the range of these quantities reduces by several orders of magnitude. The regime, on the S axis, of this steep increase coincides with that of the onset of chaotic motion in the classical dynamics. Insofar, the simple picture sketched in the Introduction is confirmed. Details of our expectation, however, need to be revised.

In particular, the notion that each splitting widens individually as the corresponding quantizing torus resolves, is not unambiguously corroborated by the data. It would imply that the transitions to a large splitting occur "from top to bottom", i.e., first for the doublet localized on the outermost torus pair within the separatrix. Indeed, if this transition is assessed from the splittings passing a certain absolute threshold, say $\Delta_\lambda = 10^{-4}$, that order is roughly obeyed. If, however, the point of onset of exponential growth, visible in a logarithmic plot like Fig. 5a, is taken as the criterion, the order is reversed.

Another remarkable fact is that the widening of the splittings and the concomitant change in character of the eigenstates, as functions of S, are continuous processes that can only roughly be associated with the decay of a KAM torus, taken as a discrete event. Even doublet states overlapping by 70% with the chaotic layer may still show a relatively small splitting and exhibit the signature of a regular state in their spatial structure and time dependence (see Fig. 6). It remains to be clarified whether this retardation of the decay of the tunnel doublets corresponds to the gradual disintegration of classical tori via cantori and vague tori.

5. Driven tunneling with dissipation

In this section, we are going to extend our working model, Eq. (1), in such a way that it allows to describe the influence of dissipation and noise on the microscopic level. We follow the usual procedure of coupling the central system to a heat bath [44, 45], by adding two terms to the Hamiltonian (1), representing, respectively, the reservoir and its interaction with the double well,

$$H(t) = H_{\text{DW}}(t) + H_{\text{I}} + H_{\text{R}}, \quad H_{\text{I}} = \sum_i x(g_i b_i + g_i^* b_i^\dagger), \quad H_{\text{R}} = \sum_i \omega_i(b_i^\dagger b_i + \frac{1}{2}). \quad (9)$$

Here, b_i, $b_i^\dagger$ are annihilation and creation operators, respectively, for a boson mode of frequency ω_i, and g_i is the corresponding coupling constant.

Proceeding in a similar way as in ref. [46], we use the density operator in the Floquet basis, reduced to the double-well degree of freedom, as the basis of our description, and resort to the usual rotating-wave and Markov approximations. This allows to derive the equation of motion for the density matrix $\tilde{\sigma}$ (in the interaction picture with respect to H_{I}), in the form of the master equation [46]

$$\dot{\tilde{\sigma}}_{\alpha,\beta}(t) = \delta_{\alpha,\beta} \sum_\nu (W_{\alpha,\nu}\tilde{\sigma}_{\nu,\nu}(t) - W_{\nu,\alpha}\tilde{\sigma}_{\alpha,\alpha}(t)) + \frac{1}{2}(1 - \delta_{\alpha,\beta}) \sum_\nu (W_{\nu,\alpha} + W_{\nu,\beta})\tilde{\sigma}_{\alpha,\beta}(t),$$

$$(10)$$

comprising a closed subset of equations for the approach of the diagonal elements towards a steady state, and another subset describing the decay of the non-diagonal elements. The coefficients $W_{\alpha,\beta}$ depend on the coupling constants and on the quasienergies; they are given elsewhere [17].

The classical limit of the quantal dynamics generated by Eq. (10) can be obtained, e.g., by switching from the density operator to a probability distribution in phase space, such as the Husimi distribution, and expanding with respect to $1/D$. Specifying the frequency dependence of the coupling strength as $|g(\omega)|^2 = \gamma\omega/\pi(1+\omega^2/\omega_c^2)$, where ω_c is a cutoff frequency, we arrive at the Langevin equation [16, 17]

$$\ddot{x} + \gamma\omega_c \int_{-\infty}^t dt'\, \dot{x}(t')\exp\left(-\omega_c(t - t')\right) - \frac{x}{2}(1 + 2\gamma\omega_c) + \frac{x^3}{16D} + S\cos\omega t = f(t). \quad (11)$$

Here, $f(t)$ is a random force with the autocorrelation function $\langle f(t)f(t')\rangle = \gamma k_{\text{B}}T\omega_c \exp(-\omega_c|t - t'|)$. Eq. (11) describes a bistable Duffing oscillator [6 – 9] with Ohmic damping and fluctuations.

The master equation (10) can now serve to investigate the influence of dissipation on the coherence effects characterizing driven tunneling, as discussed in Section 3

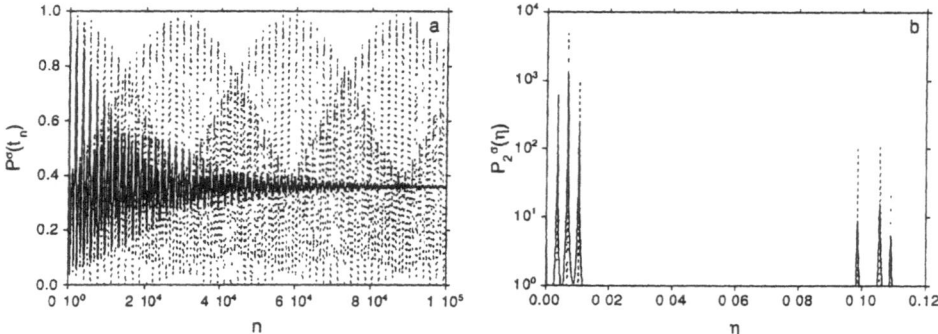

Fig. 7.   Driven tunneling with dissipation. (a) Time evolution of $P^\sigma(t_n)$ over the first 2×10^5 time steps; (b) local spectral two-point correlation function $P_2^\sigma(\eta)$ as obtained from (a). The parameter values are as in the corresponding conservative case shown in Fig. 1 (repeated here in dashed lines), but with a finite damping constant, $\gamma = 4 \times 10^{-5}$, at zero temperature.

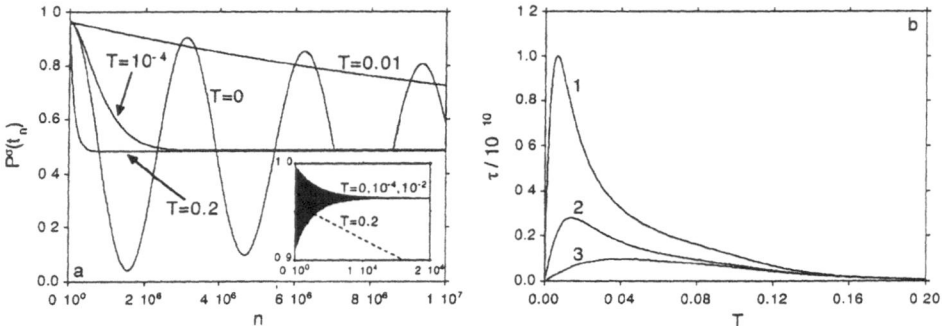

Fig. 8.   Coherent suppression of tunneling in the presence of dissipation. (a) Time evolution of $P^\sigma(t_n)$ over the first 10^7 time steps, at a parameter point ($D = 2$, $\omega = 0.01$, and $S = 3.171 \times 10^{-3}$) close to $S_{\mathrm{loc},0}(\omega)$ (see Fig. 2a), for $\gamma = 10^{-6}$ and various values of T, starting from a pure, minimum-uncertainty state centered in one of the wells (inset: the first 2×10^4 time steps on an enlarged time scale); (b) temperature dependence of the decay time τ of $P^\sigma(t_n)$ for three values of the detuning $\Delta\omega$ from the manifold $S_{\mathrm{loc},0}(\omega)$ (graph 1: $\Delta\omega = -1.4 \times 10^{-7}$, as in part (a), 2: $\Delta\omega = 5.0 \times 10^{-7}$ at $S = 3.1712 \times 10^{-3}$, 3: $\Delta\omega = 1.4 \times 10^{-6}$ at $S = 3.1715 \times 10^{-3}$). The other parameters are as in part (a). The data shown do not extend down to $T = 0$, where $\tau(T)$ diverges, but start only with the rising part of this function.

[16, 17]. Fig. 7a shows the time evolution of $P^\sigma(t_n) = \mathrm{tr}[\sigma(t_n)\sigma(0)]$ (the analogue of $P^\Phi(t_n)$, see Eq. (6)) with an initial state $\sigma(0) = |\phi(0)\rangle\langle\phi(0)|$ and parameters of H_{DW} as in Fig. 1, but with a finite damping constant $\gamma = 4 \times 10^{-5}$, at zero temperature. The complex quantum beats characteristic of the corresponding conservative system (dashed line) die out and give way to a steady state with a finite constant value of $P^\sigma(t_n)$ (in a periodically driven system, the stationary state may still possess a time dependence, with the period of the driving, which however is invisible in a stroboscopic plot like this). The broadening of the quasienergy levels,

due to the incoherent transitions described by Eq. (10), can be read off the Fourier transform of $P^\sigma(t_n)$, Fig. 7b.

A particularly interesting question is whether the coherent suppression of tunneling observed in the conservative case (see Section 3), will survive in the presence of dissipation: In order to obtain an adequate description also of this phenomenon on basis of a master equation like Eq. (10), we have to avoid part of the rotating-wave approximation used in its derivation. This approximation is valid only if the time scales of the classical relaxation and of the conservative quantal dynamics are clearly separated. However, in the vicinity of the manifolds $S_{\text{loc},k}(\omega)$ where the tunnel splitting vanishes (see Fig. 2a), exceedingly small energy scales and correspondingly large time scales occur in the undamped dynamics. This necessitates to take also quasienergy transitions into account that virtually violate energy conservation. Details of this refinement of the master equation are given in refs. [17, 47].

Fig. 8a shows the time evolution of the autocorrelation $P^\sigma(t_n)$ at a parameter point $(D = 2,\ \omega = 0.01,\ S = 3.171 \times 10^{-3})$ very close to, but not exactly on, $S_{\text{loc},0}(\omega)$, for $\gamma = 10^{-6}$ and various values of T. For low temperature, $P^\sigma(t_n)$ exhibits a slowly decaying coherent oscillation with a very long period, due to the slight offset from $S_{\text{loc},0}(\omega)$. Also here, there exist superposed oscillations reflecting the admixture of other quasienergy states. Their decay is visible only on an enlarged time scale (inset in Fig. 8a). Asymptotically, the distribution among the wells is completely thermalized. With increasing temperature, the decay time of the slow coherent oscillation first decreases until this oscillation is suppressed from the beginning (the corresponding part of the graph is not shown in Fig. 8b). After going through a minimum, however, the thermalization time increases again. At a characteristic temperature T^*, this time scale reaches a resonance-like *maximum* where the incoherent processes induced by the reservoir *stabilize* the localization of the wave packet in one of the wells and thus compensate for the detuning introduced deliberately. In Fig. 8b, we present the temperature dependence of the decay time τ (defined by $P^\sigma(t_n) \sim \exp(-n/\tau)$) for three values of the detuning $\Delta\omega$ from the manifold $S_{\text{loc},0}(\omega)$: With increasing $\Delta\omega$, the maximum is shifted towards higher temperatures and decreases in height. A variation of γ reveals that there exists a similar, resonance-like dependence also on the damping constant [47].

We emphasize that this stabilization of the coherent suppression of tunneling by noise is distinct from the trivial localization by strong damping. In fact, it has already been observed in a model simpler than the present one, where the deterministic harmonic driving of the double well was replaced by a noisy one, so that the time evolution remained unitary and a damping could not occur [12]. Rather, this phenomenon bears some resemblance to the quantum Zeno effect in a bistable system [48], and to the classical stabilization of instable equilibrium states by multiplicative noise [49, 50].

Up to now, we discussed the influence of dissipation on coherence effects. Conversely, one may also ask how the classical dynamics of the driven damped Duffing oscillator [6 – 9] is modified by quantal interference. A hallmark of the classical dynamics is the existence of attractors of various degrees of complexity, as a function of ω, S, or γ. In Figs. 9, 10 [47], we choose parameter values $(D = 6,\ S = 0.0849,\ \omega = 0.9,\ \gamma = 10^{-5}$, and $T = 0)$ where classically, there exist five limit cycles, with

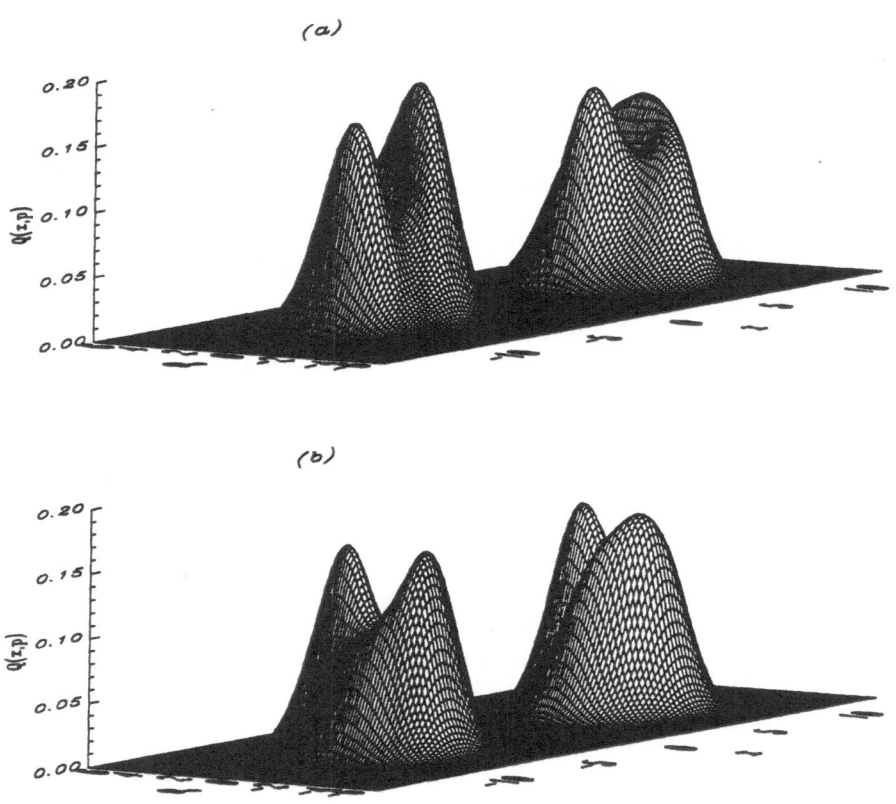

Fig. 9.   Quantal stationary state. Asymptotic Husimi distribution at $D = 6$, $S = 0.0849$, $\omega = 0.9$, $\gamma = 10^{-5}$, and $T = 0$, at phases (a) 0 and (b) $\pi/2$.

the frequency of the driving: two symmetry-related pairs with one partner within each well, and a single one encircling the wells. Fig. 9 shows the Husimi distribution in the stationary state at phases (a) 0 and (b) $\pi/2$. The broadening by quantal noise, compared to the corresponding classical, delta-like asymptotic distributions is obvious. In Fig. 10, we compare the distributions of Fig. 9a to the phase-space portrait of the corresponding conservative classical system. Both the classical attractors and the maxima of the quantal stationary distribution, while not coinciding exactly, are located near elliptic fixed points of the conservative dynamics and can be associated with the regular regions around the potential minima and the first resonance, respectively (the fifth limit cycle outside the wells is not significantly populated in the quantal stationary state). Furthermore, we see that the quantum noise preferentially broadens the stationary distribution along the limit cycle to which the corresponding

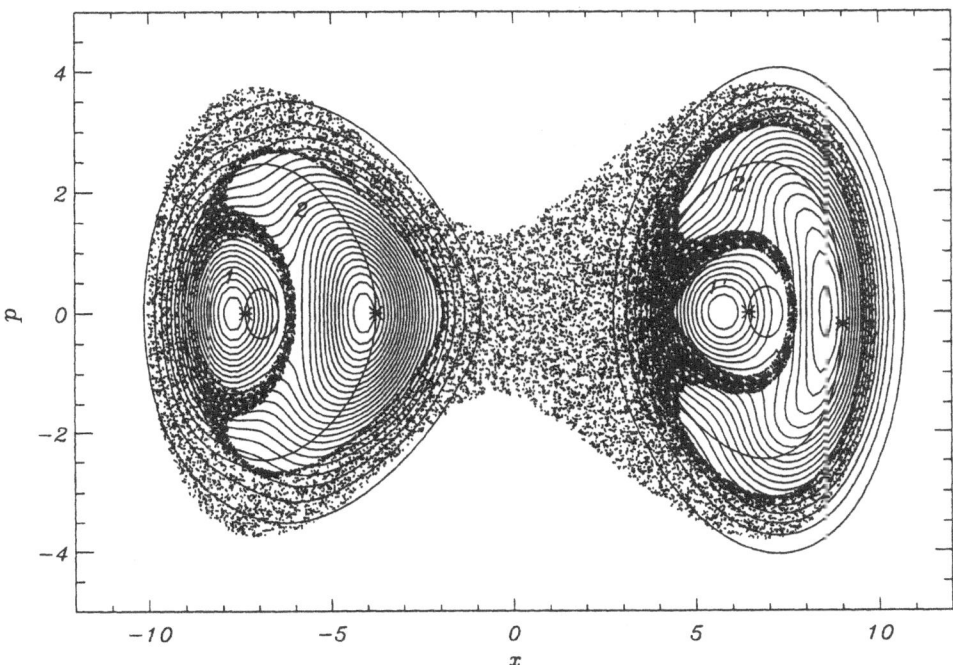

Fig. 10.   Quantal stationary state. Contour plot of the asymptotic Husimi distribution at $D = 6$, $S = 0.0849$, $\omega = 0.9$, $\gamma = 10^{-5}$, and $T = 0$, compared to the limit cycles of the corresponding classical system (the positions on the cycles at phase 0 are indicated by asterisc=), and to the phase-space portrait of the corresponding conservative dynamics.

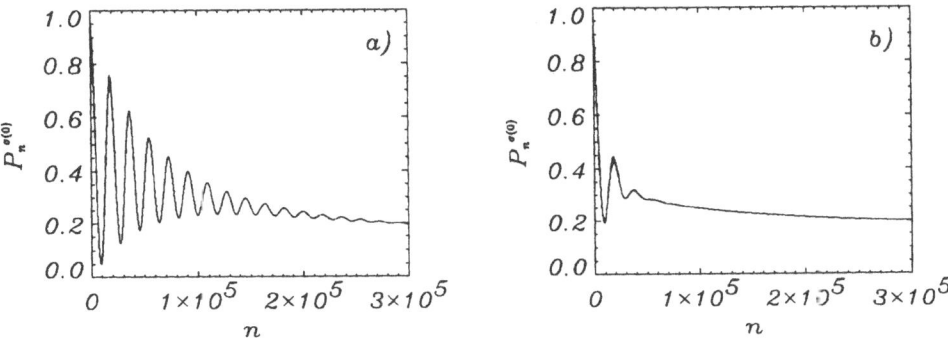

Fig. 11.   Transient tunneling between limit cycles. (a) Time evolution of $P^\sigma(t_n)$ over the first 3×10^5 time steps, at paramater values as in Figs. 9, 10, but with $\gamma = 5 \times 1C^{-6}$, for initial states prepared as coherent states located at either one of the maxima of the stationary Husimi distribution in the left well, i.e., at $p = 0$ and (a) $x = -7.5$, (b) $x = -4.2$.

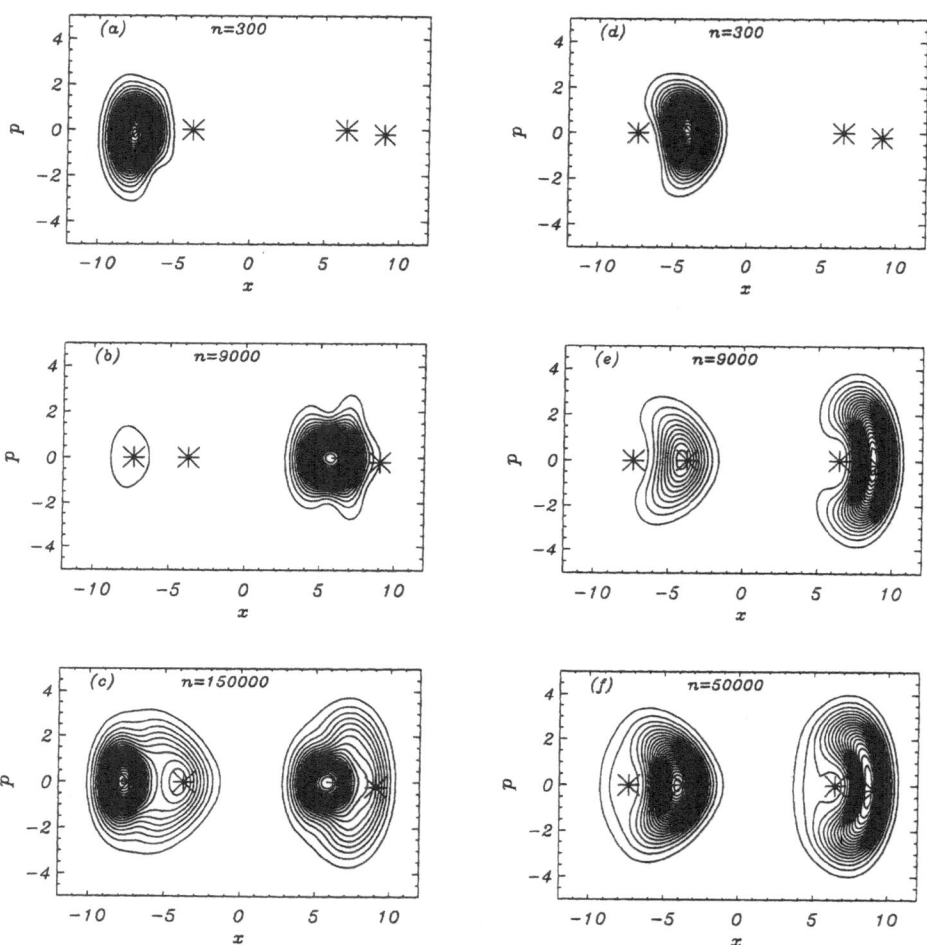

Fig. 12.   Transient tunneling between limit cycles. Husimi distribution at various times, for initial
states prepared as coherent states located at either one of the maxima of the stationary Husimi
distribution in the left well, i.e. at $p = 0$ and (a – c) $x = -7.5$, (d – f) $x = -4.2$. The positions, at
phase 0, on the limit cycles of the corresponding classical dynamics are indicated by asteriscs.

classical attractor belongs. This is the direction in which the classical phase-space
flow is least contractive [51].

While the smoothing due to quantum noise is the only quantum effect left in the
stationary state, a remnant of coherent tunneling survives in the transient behavior
[47]. Fig. 11 shows the time evolution of $P^\sigma(t_n)$ for the same parameter values as

above, with the initial states prepared as coherent states at the location of either one of the maxima of the asymptotic distribution (see Fig. 10) within the left well, corresponding to nonresonant motion (a) and to the first resonance (b), respectively. In both cases, we observe a coherent oscillation decaying as the stationary state is approached. Fig. 12 reveals that these oscillations indeed form a remnant of tunneling within each the symmetry-related pairs of regular regions. The stationary distribution among both pairs is reached only on the longer time scale of the classical relaxation. Clearly, we here observe tunneling between limit cycles.

6. Summary

The present paper is intended to highlight a number of facettes of the nonlinear dynamics in a periodically driven double-well potential, at different stages of the transition from microscopic, coherent to macroscopic, incoherent behavior. Our main tool has been a numerical analysis on basis of the Floquet formalism, which allows to speak of quasienergies and quasienergy eigenstates of the driven system, in analogy to eigenenergies and eigenstates in the undriven case.

In the deep quantal regime, we find modifications, due to the driving, of the familiar tunneling. They range from a mere acceleration of its rate, in the two extremes of slow and of fast driving, through complex quantum beats near resonances with the unperturbed system frequencies, to an almost complete suppression of tunneling by a coherent mechanism effective along one-dimensional manifolds in the parameter space spanned by amplitude and frequency of the driving.

Towards the semiclassical limit of the conservative system, the quantal behavior begins to exhibit clear traces of the classical dynamics. Specifically, we addressed the interplay between coherent transport by tunneling and diffusive transport along the chaotic layer developing in the vicinity of the separatrix of the undriven system. Eigenstate doublets residing within the paired, symmetry-related regular regions of the classical phase space exhibit exponentially small splittings and thus support tunneling. As the pair of quantizing tori pertaining to such a doublet resolves in the chaotic sea, the splitting widens and tunneling gives way to a more complex dynamics contributing to the quantal counterpart of chaotic diffusion. On a closer look, however, the scenario turns out to be less simple. For example, classical tori disintegrate only via intermediate steps, dubbed "cantori" and "vague tori", with the consequence that the transition from a regular to a chaotic nature of a quasienergy eigenstate is not sharply defined, but rather proceeds in a smooth and retarded manner. Accordingly, a strict distinction between regular and chaotic regions is inadequate on the quantum-mechanical level.

The other principal ingredient of the crossover to macroscopic behavior, besides a small relative $\hbar$, is the coherence-disturbing effect of the ambient degrees of freedom, modelled microscopically as a coupling to a quasicontinuous reservoir. As an immediate consequence, the incoherent processes render the coherence effects observed in the deep quantal regime transients. Surprisingly, however, the coherent suppression of tunneling is stabilized if damping constant and reservoir temperature are in a specific regime, a result akin to the quantum zeno effect and to the classical stabilization of instable equilibria by multiplicative noise. On the time scale of classical relaxation, the dissipative quantum dynamics approaches a stationary state which forms the analogue of the attractors of the corresponding classical dynamics. The most conspicuous quantum effect left in these stationary states is a broadening due to quantum noise. It is not isotropic, but acts preferentially in the direction where the classical phase-space flow is least contractive, that is for example, stronger along limit cycles than transverse to them.

Quite a number of questions have been left open. In the conservative case, they concern, e.g., an analytical description, in terms of semiclassical concepts, of tunneling in the presence of chaos. In the parameter regime of the dissipative system where several strange attractors coexist, both transient tunneling between their basins of attraction, and the quantal smoothing of the fractal basin boundaries [52] could be addressed. Finally, it should be checked whether the phenomenon of stochastic resonance, generated by external classical noise in periodically driven bistable systems [53], can be induced by the inherent quantum noise as well.

7. Acknowledgement

One of us (BO) acknowledges financial support by the Free State of Bavaria.

References

1. S. Chakravarty and S. Kivelson, Phys. Rev. Lett. **50**, 1811 (1983).
2. M. H. Devoret, D. Estéve, J. M. Martinis, A. Cleland, and J. Clarke, Phys. Rev. B **36**, 58 (1987); J. Clarke, A. N. Cleland, M. H. Devoret, D. Estéve, and J. M. Martinis, Science **239**, 992 (1988).
3. R. Bavli and H. Metiu, Phys. Rev. Lett. **69**, 1986 (1992).
4. J. E. Combariza, B. Just, J. Manz, and G. K. Paramonov, J. Phys. Chem. **95**, 10351 (1992).
5. L. M. Sander and H. B. Shore, Phys. Rev. B **3**, 1472 (1969).
6. P. Holmes, Philos. Trans. R. Soc. London, Ser. A **292**, 419 (1979).
7. B. A. Huberman and J. P. Crutchfield, Phys. Rev. Lett. **43**, 1743 (1979).
8. Y. Ueda, J. Stat. Phys. **20**, 181 (1979); Ann. N. Y. Acad. Sci. **357**, 422 (1980).
9. W. Szemplińska-Stupnicka, Nonlinear Dynamics **3**, 225 (1992).
10. F. Hund, Z. Phys. **43**, 803 (1927).
11. F. Grossmann, P. Jung, T. Dittrich, and P. Hänggi, Z. Phys. B **84**, 315 (1991); Phys. Rev. Lett. **67**, 516 (1991).
12. F. Grossmann, T. Dittrich, P. Jung, and P. Hänggi, J. Stat. Phys. **70**, 229 (1993).
13. F. Grossmann, T. Dittrich, and P. Hänggi, Physica B **175**, 293 (1991).
14. T. Dittrich, F. Grossmann, P. Jung, B. Oelschlägel, and P. Hänggi, Physica A **194**, 173 (1993).
15. F. Grossmann and P. Hänggi, Europhys. Lett. **18**, 571 (1992).
16. T. Dittrich, B. Oelschlägel, and P. Hänggi, Europhys. Lett. **22**, 5 (1993).
17. B. Oelschlägel, T. Dittrich, and P. Hänggi, Act. Phys. Pol. B **24**, 845 (1991).
18. R. Utermann, T. Dittrich, and P. Hänggi, in *Proceedings of the XXth International Conference on Low Temperature Physics, Eugene, 1993*, edited by R. J. Donnelly (North-Holland - Elsevier, Amsterdam, 1993).
19. R. Utermann, T. Dittrich, and P. Hänggi, submitted to Phys. Rev. E.
20. J. H. Shirley, Phys. Rev. **138B**, 979 (1965).
21. H. Sambe, Phys. Rev. A **7**, 2203 (1973).
22. N. L. Manakov, V. D. Ovsiannikov, and L. P. Rapoport, Phys. Rep. **141**, 319 (1986).
23. S. Chu, Adv. Chem. Phys. **73**, 739 (1989).
24. A. Peres, Phys. Rev. Lett. **67**, 158 (1991).
25. P. W. Anderson, Phys. Rev. **109**, 1492 (1958); Rev. Mod. Phys. **50**, 191 (1978).
26. D. R. Grempel, R. E. Prange, and S. Fishman, Phys. Rev. A **29**, 1639 (1984).
27. J. von Neumann and E. Wigner, Phys. Z. **30**, 467 (1929).

28. M. V. Berry, in *Chaotic Behavior in Quantum Systems: Theory and Applications*, edited by G. Casati, vol. 120 of *NATO ASI Series B: Physics* (Plenum, New York, 1985).
29. J. M. Gomez Llorente and J. Plata, Phys. Rev. A **45**, R6958 (1992).
30. L. E. Reichl and W. M. Zheng, in *Directions in Chaos, vol.1*, edited by H. B Lin (World Scientific, Singapore, 1987), p. 17.
31. A. J. Lichtenberg and M. A. Lieberman, *Regular and Stochastic Motion*, Vol. 38 of *Appl. Math. Sci.* (Springer, New York, 1983).
32. M. J. Davis and E. J. Heller, J. Chem. Phys. **75**, 246 (1986).
33. O. Bohigas, S. Tomsovic, and D. Ullmo, Phys. Rev. Lett. **64**, 1479 (1990); *ibid.* **55**, 5 (1990).
34. S. Tomsovic and D. Ullmo, *Tunneling in the Presence of Chaos*, preprint, 1990.
35. O. Bohigas, S. Tomsovic, and D. Ullmo, Phys. Rep. **223**, 43 (1993).
36. W. A. Lin and L. E. Ballentine, Phys. Rev. Lett. **65**, 2927 (1990); Phys. Rev A **45**, 3637 (1992).
37. J. Plata and J. M. Gomez Llorente, J. Phys. A **25**, L303 (1992).
38. H. P. Breuer and M. Holthaus, Ann. Phys. (N. Y.) **211**, 249 (1991).
39. M. V. Berry and M. Robnik, J. Phys. A **17**, 2413 (1984).
40. L. E. Reichl, in *The Transition to Chaos: In Conservative and Classical Systems: Quantum Manifestations* (Springer, New York, 1992), Chaps. 3.9, 9.5.1, and refs. therein.
41. W. P. Reinhardt, J. Phys. Chem. **86**, 2158 (1982); R. B. Shirts and W. P. Reinhardt, J. Chem. Phys. **77**, 5204 (1982).
42. K. Husimi, Proc. Phys. Math. Soc. Jap. **22**, 264 (1940).
43. R. J. Glauber, in *Quantum Optics*, edited by A. Maitland (Academic, London, 1970).
44. H. Haken, vol. XXV/2c of *Encyclopedia of Physics*, edited by S. Flügge (Springer, Berlin, 1970).
45. W. H. Louisell, *Quantum Statistical Properties of Radiation* (Wiley, London, 1973).
46. R. Blümel, A. Buchleitner, R. Graham, L. Sirko, U. Smilansky, and H. Walther, Phys. Rev. A **44**, 4521 (1991).
47. B. Oelschlägel, Ph. D. thesis, University of Augsburg, unpublished (1993).
48. M. J. Gagen, H. M. Wiseman, and G. J. Milburn, Phys. Rev. A **48**, 132 (1993)
49. R. Graham and A. Schenzle, Phys. Rev. A **26**, 1676 (1982).
50. M. Lücke and F. Schank, Phys. Rev. Lett. **54**, 1465 (1985).
51. T. Dittrich and R. Graham, Europhys. Lett. **4**, 263 (1987); Ann. Phys. (N. Y.) **200**, 363 (1990).
52. C. Grebogi, S. W. McDonald, E. Ott, and J. A. Yorke, Physics Letters **99A**, 415 (1983); S. W. McDonald, C. Grebogi, E. Ott, and J. A. Yorke, Physica D **17**, 125 (1985).
53. F. Moss, Ber. Bunsenges. Phys. Chem. **95**, 303 (1991), and refs. therein.

THE EHRENFEST THEOREM FOR MARKOV DIFFUSIONS

P. GARBACZEWSKI
Institute of Theoretical Physics
University of Wrocław, pl. M. Borna 9
PL-50 205 Wrocław, Poland

Abstract. The transformation connecting transition densities of the diffusion process with the respective Feynman-Kac kernels, induces the local field of accelerations which equals the gradient of the Feynman-Kac potential and becomes the straightforward analog of the Ehrenfest theorem.

Let us consider[1,2] a Markovian diffusion $X(t)$ in R^1 (space dimension one is chosen for simplicity) confined to the time interval $t \in [0, T]$, with the point of origin $X(0) = x_0$. The individual (most likely, sample) particle dynamics is symbolically encoded in the Itô stochastic differential equation, which we choose in the form:

$$dX(t) = b(X(t), t)dt + \sqrt{2D}\, dW(t) \qquad (1)$$

with $X(0) = x_0$, D a diffusion coefficient, $W(t)$ a normalised Wiener noise, and the drift field $b(x, t)$ is assumed to guarantee the existence and uniqueness of solutions $X(t)$. They are then non-explosive i.e. the sample paths of the process cannot escape to spatial infinity in a finite time. The rules of Itô stochastic calculus imply that the transition probability density of the process (its law of random displacements) $p(y, s, x, t), s \leq t$ solves the Fokker-Planck equation with respect to x, t

$$\partial_t p = D\Delta_x p - \nabla_x(bp) \qquad (2)$$

$$lim_{t \to s} p(y, s, x, t) = \delta(x - y) \qquad s \leq t$$

Following Stratonovich,[3] let us transform (2) by means of a substitution

$$p(y, s, x, t) = h(y, s, x, t)\frac{exp\Phi(y, s)}{exp\Phi(x, t)} \qquad (3)$$

which under the assumption that $b(x, t)$ is the gradient field

$$b(x, t) = -2D\nabla\Phi(x, t) \Rightarrow \frac{1}{2}[\frac{b^2}{2D} + \nabla b] = D[(\nabla\Phi)^2 - \Delta\Phi]\ . \qquad (4)$$

This allows us to replace (2) by the generalised diffusion equation

$$\partial_t h = D\Delta_x h - (-\partial_t\Phi + D[-\Delta\Phi + (\nabla\Phi)^2])h\ . \qquad (5)$$

$$lim_{t \to s} h(y, s, x, t) = \delta(x - y)$$

57

Z. Haba et al. (eds.), Stochasticity and Quantum Chaos, 57–62.

Its (to be strict positive) solution can be represented in terms of the Feynman-Kac (Cameron-Martin) formula, which integrates the $exp[-\int_s^t \Omega(x,u)du/2mD]$ contributions from the *auxiliary* potential $\Omega(x,t)$

$$\frac{\Omega}{m} = 2D(-\partial_t \Phi + D[-\triangle \Phi + (\nabla \Phi)^2]) = -2D\partial_t \Phi + D\nabla b + \frac{1}{2}b^2 \qquad (6)$$

with respect to the conditional[4] Wiener measure

$$h(y,s,x,t) = \int exp[-\frac{1}{2mD} \int_s^t \Omega(x,u)du]dW[y|x] . \qquad (7)$$

Since, as a consequence of (1), (2), $h(y,s,x,t)$ must be strictly positive, we recognize it as the integral kernel of the dynamical semigroup operator $exp[-\frac{1}{2mD}\int_s^t (2mD^2\triangle - \Omega)du]$ with the appropriate restrictions (continuity, boundedness from below) on $\Omega(x,t)$, and hence Φ implicit. All this is valid under the assumption that the process respects the natural[16] boundary data where the density of the diffusion (hitherto not explicitly introduced) vanishes, with boundary points at infinity.

Given $p(y,s,x,t)$, we can utilise the Itô formula [1,2,5,8] which for any smooth function of the random variable states that its forward time derivative in the conditional mean, reads

$$lim_{\triangle t \downarrow 0} \frac{1}{\triangle t}[\int p(x,t,y,t+\triangle t)f(y,t+\triangle t)dy - f(x,t)] = . \qquad (8)$$

$$(D_+f)(X(t),t) = (\partial_t + b\nabla + D\triangle)f(X(t),t)$$

with $X(t) = x$. Then, for the second forward derivative (in the conditional mean) of the diffusion process $X(t)$, in virtue of (4), (6) we have

$$(D_+^2 X)(t) = (D_+b)(X(t),t) = (\partial_t b + b\nabla b + D\triangle b)(X(t),t) = \frac{1}{m}\nabla \Omega(X)t),t) \qquad (9)$$

This formula is a precise embodiment of the second Newton law (in the conditional mean) governing *all* Markovian diffusions consistent with (1)-(7), albeit it is "Euclidean looking". The *auxiliary potential* $\Omega(x,t)$ *plays here the role of the corresponding force field potential*: a bit surprising outcome for anyone familiar with the large friction (Smoluchowski) limit of the phase space Brownian motion, however definitely[15] an inevitable one

Our previous discussion refers to the individual (sample) features of a particle propagation in contact with the randomly perturbing environment: the Wiener noise is superimposed on the systematic field $b(x,t)$ of local drifts. By attributing an initial probability distribution $\rho_0(x) = \rho(x,0)$ to the random variable $X(t)$, we pass to the *statistical ensemble* (hence collective) analysis. Because of (1), (2) the forward dynamics of the density $\rho(x,t) = \int \rho_0(y)p(y,0,x,t)dy$ is uniquely defined. The microscopic law of random displacements $p(y,s,x,t), s \leq t$ generates all possible random propagation scenarios (sample paths) from each chosen point of origin $X(0) = x_0$, for the flight duration times $t > 0$. The statistical outcome (prediction about the most likely future of an individual particle) is casually considered as independent of the assumed probability distribution $\rho(x_0)$. However, once introduced this density sets a statistical correlation between individual members of the ensemble, even if there are no mutual interactions to be accounted for. An interesting *ensemble* characterisation of the random motion is here possible by introducing (for Markov processes only) the transition density $p_*(y, s, x, t)$

$$\rho(x,t)p_*(y,s,x,t) = p(y,s,x,t)\rho(y,s) \tag{10}$$

which allows to trace back the most likely statistical past of particles *conditioned to comprise the evolving statistical ensemble* with the distribution $\rho(x,t)$. One should consult Refs. 6,7 to realize that any realistic diffusion (free Brownian motion included !) admits (10): it has nothing to do with a physically realizable reversal of the generally irreversible process. In this case[5,8] we can define the *backward* time derivative of the process $X(t)$ (now supplemented by the distribution $\rho(x,t)$), which in the jointly conditional and ensemble[6,7] mean reads:

$$lim_{\Delta t \downarrow 0} \frac{1}{\Delta t}[x - \int p_*(y, t - \Delta t, x, t)y dy] = (D_- X)(t) = b_*(X(t), t) \tag{11}$$

with the corresponding Itô formula for $f(x,t)$

$$(D_- f)(X(t),t) = (\partial_t + b_* \nabla - D\Delta)f(X(t),t) \tag{12}$$

Because of (10) the drifts $b(x,t)$ and $b_*(x,t)$ are *not* mutually independent, and indeed[5,8,9] on domains free of nodes (ρ vanishing at the boundaries) we have

$$b_*(x,t) = b(x,t) - 2D\nabla ln\rho(x,t) . \tag{13}$$

Consequently, the current velocity[5] field

$$v(x,t) = \frac{1}{2}(b + b_*)(x,t) \tag{14}$$

can be viewed as the supplementary to $\rho(x,t)$ (it induces the osmotic velocity[5] notion $u(x,t) = D\nabla ln\rho(x,t) = \frac{1}{2}(b - b_*)$ in turn) characteristic of the stochastic flows. This time, elevated to the macroscopic (statistical ensemble) level. In terms of the local velocity fields $u(x,t), v(x,t)$ both of which are gradient fields, one can explicitly[10-12] demonstrate that

$$(D_+^2 X)(t) = \partial_t v + v\nabla v + \frac{1}{m}\nabla Q = (D_-^2 X)(t) \tag{15}$$

$$Q(x,t) = 2mD^2 \frac{\Delta\rho^{1/2}}{\rho^{1/2}}$$

which extends the identity (9) to $(D_-^2 X)(t)$. With the density $\rho(x,t)$ in hand, we can evaluate the mean (ensemble expectation) values of (15) and (9)

$$E[(D_+^2 X)(t)] = E[(D_-^2 X)(t)] = \frac{1}{m}E[\nabla\Omega(X(t),t)] \tag{16}$$

where because of (cf. the original version of the Ehrenfest theorem[13,14] in quantum mechanics, which exploits the previously mentioned property that the probability density vanishes at the boundaries of the integration volume

$$E[\nabla Q(X(t), t)] = 0 \tag{17}$$

there holds a classical Liouville equation in the mean, with the "Euclidean looking" potential (in view of the minus sign absence)

$$E[(\partial_t v + v\nabla v)(X(t), t)] = \frac{1}{m} E[(\nabla \Omega)(X(t), t)] . \tag{18}$$

On the other hand, in virtue of the continuity equation, we have

$$E[X(t)] = \int x\rho(x, t) dx \quad \Rightarrow$$

$$\frac{d}{dt} E[X(t)] = \frac{1}{2}(E[D_+X] + E[D_-X]) = E[v(X(t), t)] \tag{19}$$

and furthermore (see also Ref. 15)

$$\frac{d^2}{dt^2} E[X(t)] = \frac{d}{dt} E[v(X(t), t)] = E[(\partial_t v + v\nabla v)(X(t), t)] = \frac{1}{m} E[\nabla \Omega(X(t), t)] \tag{20}$$

hence the "Euclidean looking " second Newton law is found to be respected by the diffusion process (1) both in the conditional (9) and the ensemble (15), (20) mean.

Notice that the auxiliary potential in the form $\Omega = 2Q - V$ where V is any Rellich class (to allow for the Feynman-Kac formula for the semigroup kernel) representative, defines drifts of Nelson's diffusions, for which $E[\nabla Q] = 0 \Rightarrow E[\nabla \Omega] = -E[\nabla V]$ i.e. the "standard looking" form of the second Newton law in the mean arises.

Our previous discussion associates an a priori given drift (control) field $b(x, t), t \in [0, T]$ with a potential $\Omega(x, t)$. Clearly, we encounter here a fundamental problem of what is to be interpreted by a physicist (external observer) as the external force field manifestation in the diffusion process. Let us invert our previous reasoning and take not $b(x, t)$ but $\Omega(x, t), t \in [0, T]$ to be given a priori as a *primary dynamical control* for the Markovian diffusion (1), (2), which we are in principle capable of manipulating (the role attributed to the external observer). Then, we shall say that the diffusion respects the second Newton law in the conditional mean, if

$$(D_+^2 X)(t) = \frac{1}{m} \nabla \Omega(X(t), t) \tag{21}$$

holds true.

The evolution in time of the gradient drift field $b(x, t)$ and this (given a priori) of $\Omega(x, t)$ are *compatible* if

$$\partial_t b + b\nabla b + D\Delta b = \frac{1}{m} \nabla \Omega \tag{22}$$

$$b_0(x) = b(x, 0) .$$

It is a *sufficient* compatibility condition, which allows us to derive the drift dynamics from this $\Omega(x, t)$. In the time-independent case there is no real freedom in the choice of the initial Cauchy data for Eq. (22), and an identity $\Omega_0(x) = m(D\nabla b_0 + \frac{1}{2}b_0^2)(x) = \Omega(x, 0)$ must be satisfied.

Eq. (22) sets a well defined Cauchy problem for $b(x, t)$ in terms of $\Omega(x, t)$. If we associate an initial probability distribution $\rho_0(x)$ with $X(0)$, then our (sufficient) compatibility condition (22) can be *equivalently (!)* written as the coupled Cauchy problem

$$\partial_t \rho = -\nabla(\rho v)$$

$$\partial_t v + v \nabla v = \frac{1}{m} \nabla(\Omega - Q) \tag{23}$$

$$\rho_0(x) = \rho(x, 0), v_0(x) = v(x, 0)$$

where $b_0(x) = v_0(x) + D\nabla ln\rho_0(x)$, with the initial data essentially unrestricted, except for the time-independent case.

Remark 1: One should not be misled by the seemingly complicated form of the nonlinear coupled Cauchy problem (23). It is precisely Eq. (22), which guarantees its solvability. Indeed, by virtue of the standard path integral identity[1]:

$$p(y, s, x, t) = lim_{\Delta t \downarrow 0} \int dz_1 ... \int dz_n (4\pi D\Delta t)^{-n/2} \tag{24}$$

$$exp(-\frac{1}{4\pi D\Delta t} \sum_{k=0}^{n-1} [z_{k+1} - z_k - b(z_k, t_k)\Delta t]^2)$$

$$\Delta t = \frac{t - s}{n}, z_0 = y, z_n = x, t_0 = s, t_n = t$$

it suffices to know the time development of the drift $b(x, t)$ to have uniquely specified the time evolution of $\rho(x, t) = \int p(y, s, x, t)\rho(y, s)dy$, once $\rho_0(x)$ is given.

Remark 2: Since

$$p(y, s, x, t) = lim_{\Delta t \downarrow 0} \int dz_1 ... \int dz_n \prod_{k=o}^{n-1} p(z_k, t_k, z_{k+1}, t_{k+1}) \tag{25}$$

we can perform the Stratonovich substitution (3) for each entry separately, and observe[3] that

$$p(y, s, x, t) = exp[\Phi(y, s) - \Phi(x, t)] lim_{\Delta t \downarrow 0} \int dz_1 ... \int dz_n \prod_{k=0}^{n-1} h(z_k, t_k, z_{k+1}, t_{k+1}) \cdot \tag{26}$$

The semigroup composition property is here clearly seen. It in turn justifies the procedures of Refs. 10–12, see also Refs. 15–17.

Acknowledgment

The present note is an excerpt from Ref. 16, which was completed during my stay at the University of Kaiserslautern. I am willing words of gratitude to Professor J. Kupsch for hospitality.

References

1. H. Risken, The Fokker-Planck equation, Springer, Berlin, 1989
2. Ph. Blanchard, Ph. Combe, W. Zheng, Mathematical and physical aspects of stochastic mechanics, Lect. Notes in Phys. vol. 281, Springer, Berlin, 1987
3. R.L. Stratonovich, Select. Transl. Math. Statist. and Probability, 10, (1971), 273
4. J. Glimm, A. Jaffe, Quantum physics-a functional integral point of view, Springer, Berlin, 1987
5. E. Nelson, Quantum fluctuations, Princeton Univ. Press., Princeton, 1985
6. P. Garbaczewski, Phys. Lett. A 162, (1992), 129
7. P. Garbaczewski, J.P. Vigier, Phys. Rev. A 46, (1992), 4634
8. F. Guerra, Phys. Reports, 77, (1981), 263
9. M. Kac, Probability and related topics in physical sciences, Interscience Publ., NY, 1959
10. J.C. Zambrini, Phys. Rev. A 33, (1986), 1532
11. J.C. Zambrini, J. Math. Phys. 27, (1986), 2307
12. P. Garbaczewski, Phys. Lett. A 172, (1993), 208
13. H.E. Wilhelm, Phys. Rev. D 1, (1970), 2278
14. P. Ehrenfest, Z. Physik, 45, (1927), 455
15. P. Garbaczewski, Phys. Lett. A 178, (1993), 7
16. P. Garbaczewski, Relative Wiener Noises and the Schrödinger Dynamics in External Force Fields, Univ. Kaiserslautern preprint KL–Th–93/12
17. Ph. Blanchard, P. Garbaczewski, Natural boundaries for the Smoluchowski equation and affiliated diffusions, to appear in Phys. Rev. E

THE QUANTUM STATE DIFFUSION MODEL, ASYMPTOTIC SOLUTIONS, THERMAL EQUILIBRIUM AND HEISENBERG PICTURE

N. GISIN

University of Geneva, GAP, 1211 Geneva 4, Switzerland
Fax: +41 − 22 − 781.0980, e-mail: Gisin@sc2a.unige.ch

Abstract. The Quantum State Diffusion (QSD) model is summarized. A criterion determining the asymptotic solution is presented and illustrated for the case of a harmonic oscillator in a thermal bath. The QSD Heisenberg picture and multi-time expectation values are also summarized. First results on the non-Markovian case are presented. Finally the question of whether QSD is mathematics or physics is addressed.

1. Introduction

The dynamics of open quantum systems is described, in the Markovian limit, by master equations for the density operator ρ :

$$\frac{d\rho_t}{dt} = [\mathcal{L}]\rho_t = -i[H,\rho_t] + \sum_j \left(L_j\rho_t L_j^+ - \frac{1}{2}\left\{ L_j^+ L_j, \rho_t \right\} \right) \tag{1}$$

where $[\mathcal{L}]$ is the Liouville (super) operator, H is the Hamiltonian of the system and the L_j are linear operators that take into account the effects of the environment. The density operator ρ_t contains all the information about single time expectation values. These expectation values are measured in actual quantum experiments by repeating a measurement on individual systems and then computing the average. The Quantum State Diffusion (QSD) model [1] associates in a canonical way to each master equation (1) a pure state valued stochastic process such that the expectation value in the model are computed by averaging over many "runs" (i.e. realization), as in the real experiments. In this way one gains over the traditional approach to open quantum systems:

1. →the possibility to represent individual runs of experiments,
2. →the possibility to compute numbers, directly related to experimental results, using only state vectors instead of density matrices, gaining thus computational power,
3. → a quantum measurement is nothing special, it is just one example of the interaction of a system with its environment,
4. → a nice mathematical example of stochastic processes in Hilbert spaces.

The stochastic process that defines the QSD model is determined by the following (Itô) stochastic equation [1,2]:

63

Z. Haba et al. (eds.), Stochasticity and Quantum Chaos, 63–71.

$$d\psi_t = -iH\psi_t \, dt + \frac{1}{2}\sum_j \left(2 < L_j^+ >_{\psi_t} L_j - L_j^+ L_j - < L_j >_{\psi_t} < L_j^+ >_{\psi_t}\right)\psi_t \, dt$$

$$+\sum_j (L_j - < L_j >_{\psi_t})\,\psi_t \, d\xi_{jt}$$

(2)

where the ξ_{jt} are complex-valued Wiener processes of zero mean and correlations:

$$d\xi_{it}\, d\xi_{jt} = 0, \quad d\xi_{it}\, d\xi_{jt}^* = \delta_{ij}dt.$$

(3)

Eq. (1) and (2) are related by averaging the pure states ψ_t over the noises ξ_{jt}. Indeed, from (2) one deduces that $d\,|\psi_t| = 0$ and:

$$d\psi_t\, \psi_t^+ = [\mathcal{L}]\psi_t\, \psi_t^+ dt+$$

$$\sum_j (L- < L_j >_{\psi_t})\,\psi_t\, \psi_t^+ d\xi_{jt} \; + \psi_t\, \psi_t^+ \sum_j (L- < L_j >_{\psi_t})\, d\xi_{jt} \; .$$

Hence: $\ll \psi_t\, \psi_t^+ \gg_\xi = \rho_t$.

In [1] we emphasized the use of eq. (2) for practical computations based on a Monte Carlo type algorithm, specially in quantum optics. We also emphasize the remarkable convergence of two trends in physics that have previously been quite distinct: The measurement "problem" as considered by physicists worried about the foundations of quantum physics and the quantum measurement process as treated pragmatically by experimenters looking for intuitive pictures and rules for computations. In [3] we proved general localization theorems and in [4] we emphasized the physical picture and insight provided by the state diffusion model. In [5] the model is used to describe a quantum jump experiment. Some results about existence of solutions of $eq(2)$ in the infinite dimensional case can be found in [6]. Finally, reference [7] contains some explicit solutions.

Equation (2) is nonlinear. However, rewriting it in Stratonovich form one gets:

$$d\psi_t = -iH\psi_t \, dt - \frac{1}{2}\sum_j \left(L_j^+ L_j - < L_j^+ L_j >_{\psi_t}\right)\psi_t dt$$

$$+\sum_j (L- < L_j^+ >_{\psi_t})\,\psi_t \circ \left(d\xi_{jt} + < L_j^+ >_{\psi_t} dt\right)$$

one can thus realize that there are two kinds of nonlinearities: the mean values subtracted from the operator, necessary only for the normalization of the vector ψ_t, and the shift of the noise term $d\xi_j$. The latter represents the back action of the system on its environment.

2. Asymptotic Solutions

Asymptotic solutions of eq (2) can be found in [6,11]. The dissipating terms will drive the system toward regions of the Hilbert space where the fluctuations are minimal:

$$Min \left(\sum_j ((L_j - <L_j>_{\psi_t}) \psi_t \, d\xi_{jt} \right)^2 = Min \sum_j (\Delta L_j)^2 . \qquad (4)$$

If the environment operator L is self-adjoint these minima correspond to the eigenstates. In this way one can model idealized measurements, reproducing the projection postulate as asymptotic solutions: $\psi_0 \overset{t \to \infty}{\longrightarrow} \varphi_n$ with probability $|<\varphi_n|\psi_0>|^2$ where φ_n is an eigenstate of L [1,8].

If $L = a$ or a^+, the annihilation or creation operator of the harmonic oscillator, these minima correspond to the coherent states. This example is developed in the next section.

Note that these states of minimum fluctuations are also those satisfying Zurek conditions [12] in that they remain maximally pure under the evolution (1):

$$Min \sum_j (\Delta L_j)^2 = Max \ \ Tr \left(P_{\psi_t} \frac{d}{dt} P_{\psi_t} \right) = Max \ \frac{d}{dt} Tr \left(P_{\psi_t}^2 \right)$$

The Quantum State Diffusion model provides thus explicit equations for the "narrow wave packets" discussed, for instance, by Zurek [12] and Zeh [13]. Note that if the states minimizing $\sum_j (\Delta L_j)^2$ are not preserved by the Hamiltonian, then the asymptotic solutions of the quantum state diffusion model (2) are no longer given by (4), but result from a competition between the Hamiltonian and dissipative part of the evolution equation, as described by (2). This contrasts with the Zurek condition, since any Hamiltonian preserves pure states.

3. Thermal equilibrium

The master equation describing the approach to thermal equilibrium of a harmonic oscillator is [14]:

$$\frac{d\rho_t}{dt} = - i\omega \left[a^+ + a, \rho_t \right] +$$

$$\bar{n} \, \gamma \left(a^+ \rho_t a - \frac{1}{2} \{aa^+, \rho_t\} \right) + (\bar{n} + 1) \, \gamma \left(a \rho_t a^+ - \frac{1}{2} \{a^+ a, \rho_t\} \right) \qquad (5)$$

where $\bar{n}$ is the equilibrium mean photon number, $\bar{n} = <a^+ a>_{equ}$, and γ^{-1} is the relaxation time. The corresponding QSD model has been considered in [15].

Any initial state evolves asymptotically under the stochastic equation (2) corresponding to (5) towards a coherent state of the harmonic oscillator. The complex number α that labels the coherent states $\mid \alpha > (a \mid \alpha >= \alpha \mid \alpha >)$ follows then the well known diffusion equation:

$$d\alpha_t = i\omega\alpha_t \, dt - \frac{\gamma}{2}\alpha_t \, dt + (\bar{n} \, \gamma)^{1/2} \, d\xi_t \ .$$

This examples illustrates two general feature of QSD [2,7,16]:

1. Any initial states localizes (with probability one). The localization may be on energy eigenstates, or in position or in phase space, depending on the environment operators.

2. After localization, the wavepacket spread is the minimum compatible with the Heisenberg uncertainty relations (quantum fluctuations) and the wavepacket center follows a classical trajectories with thermal fluctuations. At zero temperature, $\bar{n} = 0$, and the trajectory is deterministic.

4. Heisenberg picture of the QSD model [17]

Since the quantum state diffusion equation (2) is nonlinear, the corresponding Heisenberg picture maps linear initial operator A_0 to "nonlinear operator" A_t. The latter are defined as maps associating to any pair of bra and ket a complex number:

$$A_t \ : \mathcal{H}^* \ \times \ \mathcal{H} \to \mathbb{C} \ .$$

The Heisenberg picture of the operator A corresponding to the quantum state diffusion (1) and (3) is given by:

$$A_t \ (\varphi, \psi) \ = < \varphi_t \mid A \mid \psi_t >$$

where φ_t and ψ_t are the solution of a pair coupled of stochastic eqs. with initial conditions φ and ψ, respectively. This pair of stochastic equations is chosen such that:

a. $A_t \ (\psi, \psi) \ = < A >_{\psi_t}$, as it should be for the relation between the Schrödinger and Heisenberg pictures, and

b. $\mathbf{1}_t = \mathbf{1}$, i.e. the unit operator is preserved.

From these two conditions one finds:

$$d\psi_t = -iH\psi_t \, dt + \frac{1}{2}\sum_j \left(2\ell_j \, (\psi_t, \varphi_t)^* \, L_j - L_j^+ L_j - \ell_j(\varphi_t, \psi_t)\ell_j \, (\psi_t, \varphi_t)^*\right) \psi_t \, dt$$

$$+ \sum_j (L_j - \ell_j(\varphi_t, \psi_t))\psi_t \, d\xi_{jt}$$

$$(6a)$$

$$d\varphi_t = -iH\varphi_t \, dt + \frac{1}{2}\sum_j \left(2\ell_j \left(\varphi_t, \psi_t\right)^* L_j - L_j^+ L_j - \ell_j \left(\psi_t, \varphi_t\right) \ell_j \left(\varphi_t, \psi_t\right)^*\right) \varphi_t \, dt$$

$$+ \sum_j (L_j - \ell_j(\psi_t, \varphi_t))\varphi_t \, d\xi_{jt}$$

(6b)

where $\ell_j(\alpha, \beta) = <\alpha |L_j|\beta> / <\alpha | \beta>$ and the ξ_{jt} are complex-valued Wiener processes of zero mean and correlations given by (3). From eqs (6) it follows that $d<\varphi_t | \psi_t> = 0$, hence nonorthogonal vectors remain nonorthogonal, and condition b is satisfied. For $\varphi = \psi$ the eqs. (6) reduce to (2), hence condition a is also satisfied.

The evolution equation of $A_t(\varphi, \psi)$ can be deduced from (6) with the usual Itô calculus:

$$dA_t \left(\varphi, \psi\right) = -i <\varphi_t | [A, H] | \psi_t> dt + \sum_j <\varphi_t | L_j^+ A L_j - \frac{1}{2}\left\{L_j^+ L_j, A\right\} | \psi_t> dt$$

$$+ <\varphi_t | A \sum_j (L_j - \ell_j(\varphi_t, \psi_t)) | \psi_t> d\xi_{jt}$$

$$+ <\varphi_t | \sum_j \left(L_j^+ - \ell_j(\psi_t, \varphi_t)^*\right) A | \psi_t> d\xi_{jt}^* \, .$$

This gives the Heisenberg form of the quantum state diffusion equation. Accordingly, the mean of the "nonlinear operator" A_t over the noises ξ_j is equal to the usual linear Heisenberg operator A_t^H :

$$M A_t(\varphi, \psi) = <\varphi | A_t^H | \psi> \, .$$

The Heisenberg picture of QSD can be used as a practical tool to compute multi-time expectation values (that can not be computed from the density operator ρ_t) [17]. The two-time correlation functions, for instance, can be expressed in terms of the "nonlinear Heisenberg operators" and an arbitrary ortho-normal bases $\{\theta_j\}$ as follows:

$$M \left(\sum_j A_{t_1}(\psi, \theta_j) B_{t_2}(\theta_j, \psi)\right) = \sum_j M\left(A_{t_1}(\psi, \theta_j)\right) M\left(B_{t_2}(\theta_j, \psi)\right)$$

$$= \sum_j <\psi | A_1^H | \theta_j> <\theta_j | B_{t_2}^H | \psi>$$

$$= <\psi | A^H(t_1) B^H(t_2) | \psi>$$

which is the usual quantum correlation function $g_{AB}(t_1, t_2)$. Note that A_{t_1} and B_{t_2} are independent stochastic processes, i.e. the Wiener processes $d\xi_j$ governing A_{t_1} and B_{t_2} are independent.

5. Non-Markovian Master equations and colored noise.

In this section we present some preliminary results on non-Markovian master equations in the QSD model context. There are two immediate difficulties.

1. There is no equivalence to the Lindblad theorem classifying the master equations. Moreover, it is not obvious to check whether an evolution equation with a memory term i.e. an integro-differential equation) preserves positivity.
2. Stochastic equations with colored noise are rarely soluble.

As a first step one can consider the case of pure states evolving under a dichotomic noise $I_t = \pm 1$:

$$\frac{d\psi_t}{dt} = -i(H_0 + H_1\, I_t)\,\psi_t \ . \tag{7}$$

Let χ^{-1} be the mean time between the switches $I_t \to -I_t$. At each time the solution of (7) determines two density operators $\rho_+(t)$ and $\rho_-(t)$ that represent the average of the pure states correlated – at that time – with the value $+1$ and -1 of the noise I_t. Note that $\rho_+(t)$ and $\rho_-(t)$ are not normalized, but $Tr(\rho_+(t)) + Tr(\rho_-(t)) = 1$. These two operators evolve according to:

$$\frac{d\rho_+}{dt} = -i[H_0 + H_1, \rho_+] - \chi(\rho_+ - \rho_-)$$

$$\frac{d\rho_-}{dt} = -i[H_0 - H_1, \rho_-] + \chi(\rho_+ - \rho_-) \ .$$

Defining $\rho_t = \rho_+(t) + \rho_-(t)$ and $\sigma_t = \rho_+(t) - \rho_-(t)$ one gets:

$$\frac{d\rho_t}{dt} = -i[H_0, \rho_t] - i[H_1, \sigma_t]$$

$$\frac{d\sigma_t}{dt} = -i[H_0, \sigma_t] - i[H_1, \rho_t] - 2\chi\sigma_t \ .$$

Note that ρ_t is the density matrix of the system. It is normalized to unity (provided the initial state is normalized: $Tr(\rho_t) = 1$. The trace of σ_t, on the contrary, vanishes (assuming an even initial distribution of the noise I_0): $Tr(\sigma_t) = 0$, and $\sigma_0 = 0$.

Next, going to the interaction representation, $\tilde{\rho}_t$ and $\tilde{\sigma}_t$ one obtains:

$$\frac{d\tilde{\rho}_t}{dt} = -\left[\tilde{H}_1(t), \int_0^t \frac{\tilde{\sigma}_s}{dt}\, ds\right]$$

$$= - \left[\tilde{H}_1(t), \int_0^t \left[\tilde{H}_1(t-s), \rho_{\tilde{t}-s} \right] e^{-2\chi s} \, ds \right] . \tag{8}$$

Accordingly a quite large class of non-markovian master equations is recovered. Actually all equations arising by tracing out the environment, assuming a fixed state for the environment with exponentially decreasing correlations and with an interaction Hamiltonian of the form $A \otimes B$, are of that form [18].

In the markovian limit, eq. (8) recovers the QSD equation (with self-adjoint environment operators), but with purely imaginary noise. It corresponds thus to phase mixing without reduction (i.e. without localization).

Conclusions: is QSD Mathematics or Physics?

The Quantum State Diffusion (QSD) model [1] has been developed in the course of a research program [19..24,8,9] that aims at:

1. Overcoming the dualistic dynamics of quantum mechanics: the unitary Schrödinger evolution on one side, and the projection postulate on the other side,

2. Avoiding the need to attribute a wavefunction to the entire universe (in contrast to Everett type of interpretations). Hence, avoiding unnecessary "beings" in the theory.

In its present form the QSD model does not satisfy the above requirements, though it goes in this direction. It suffers from 2 related limitations:

1. There is still a need to divide the universe into a system and an environment,

2. The system's dynamic is assumed to be markovian.

Despite these limitations the QSD model has received quite some attention in the recent past, partly thanks to the GRW model [23] and to Bell's advertisement [24] of it (this model first appeared with jumps, but has then quickly been translated into Itô continuous stochastic processes [25,8,26]), and partly because of its relation to the Monte-Carlo wavefunction approach [27..29]. But the attention comes also - and I like to think mainly - because:

1. It provides appealing pictures in a systematic way,

2. It can be - and has been - used for Monte-Carlo computation of density matrix dynamics,

3. The stochastic wavefunctions do generically localize either on energy eigenstates, or around classical trajectories in phase space.

Now the 'big' question: is QSD only Mathematics or does it contain some Physics? The answer will only be clear in the future. But it is already clear that there is a dichotomic choise: either the Schrödinger equation is all there is and each measurement like interaction splits the Universal wave function into a superposition of word-components; or only one world exists and the Schrödinger equation is only a (good) approximation. For most physicist the first alternative is obvious: Weinberg [30], in his book "Dreams of a final theory", argues even that since quantum mechanics is hard to modify [31] it must be part of the final theory of Physics! Nevertheless, one may apply Ockham's razor argument and note that the assumption of a universal validity of the Schrödinger equation implies an exponentially increasing infinity of unknown and untestable facts (all the other world-components). Zeh [32] has responded to that argument by applying Ockham's razor to the laws, rather than to the facts....make up your choice! But notice that:

1. The Schrödinger equation is no longer the best For All Practical Purposes [32,33]!

2. In practice physicsts never use the universal wavefunction, but only the component that is relevant for their purposes. Hence the Schrödinger equation, in practice, is anyway never used without a cut.

3. New physics is more likely to emerge from new theories than from old ones!

4. Why should a theory well-known for its stochasticity be based on a deterministic equation [34]?

Acknowledgments

It is a great pleasure to acknowledge stimulating discussions with Ian Percival, Lajos Diosi, J. Halliwel and D. Zeh.

References

1. N. Gisin and I.C. Percival, J. Phys. **A25**, 5677, 1992; see also Phys. Lett. **A167**, 315, 1992.
2. N. Gisin and M.B. Cibils, J. Phys. **A25**, 5165, 1992.
3. N. Gisin and I.C. Percival, J. Phys. **A26**, 2233, 1993.
4. N. Gisin and I.C. Percival, J. Phys. **A26**, 2245, 1993.
5. N. Gisin, P.L. Knight, I.C. Percival, R.C. Thompson and D.C. Wilson, "Quantum state diffusion theory and a quantum jump experiment", J. Modern Optics, in press, 1993.
6. D. Gatarek and N. Gisin, J. Math. Phys. **32**, 2152–2157, 1991.
7. Y. Salama and N. Gisin, Phys. Lett. **A**, in press, 1993.
8. N. Gisin, Phys. Rev. Letts. **52**, 1657–1660, 1984; and Helv. Phys. Acta **62**, 363–371, 1989.
9. L. Diosi, Phys. Lett. **129A**, 419, 1988; and J. Phys. **A21**, 2885–2898, 1988.
10. D. Chruscinski and Staszewski, Phys. Scr. **45**, 193, 1992.
11. A. Barchielli and V.P. Belavkin, J. Phys. **A24**, 1495, 1991.
12. W.H. Zurek, Physics Today, p 36 October 1991; and p 13 April 1993; Phys. Rev. Lett. **70**, 1187, 1993.
13. H.D. Zeh, Phys. Lett. **A172**, 189, 1993, see also N. Gisin and I. Percival, Phys. Lett. **A175**, 144, 1993.
14. F. Haake, "Statistical treatment of open systems by generalized master equations", Vol. 66 of Springer Tracts in Modern Physics, 1973.
15. T.P. Spiller, B.M. Garraway and I.C. Percival, Phys. Lett. **A179**, 63, 1993.
16. I.C. Percival, "Localisation of wide open quantum systems", preprint, 1993.
17. N. Gisin, "Time correlations and Heisenberg picture in the quantum state diffusion model of open systems", J. Modern Optics, in press, 1993.
18. E.B. Davies, "Quantum Theory of Open Systems", Academic Press, London, 1976.
19. D. Bohm and J. Bub, Rev. Mod. Phys. **38**, 473–475, 1966.
20. F. Karolyhazy, Nuovo Cim. **52**, 390, 1966.
21. P. Pearle, Phys. Rev. D **13**, 857–868, 1976, & Internat. J. Theor. Phys. **18**, 489–518, 1979, & J. Stat. Phys. **41**, 719–727, 1985.
22. A. Shimony, "Desiderata for a modified quantum dynamics" in: Philosophy of Science Association, 1991.
23. G-C. Ghirardi, A. Rimini and T. Weber, Phys. Rev. **D34**, 470–491, 1986.
24. J. S. Bell, in Schrödinger, Centenary of a polymath, Cambridge Uni. Press, 1987.
25. P. Pearle, Phys. Rev. **A39**, 2277, 1989.
26. G-C. Ghirardi, Ph. Pearle and A. Rimini, Phys. Rev. **42A**, 78, 1990.
27. H.J. Carmichael, "An open systems approach to quantum optics", ULB Lectures in Nonlinear Optics, in press, 1993.

28. J. Dalibard, Y. Castin and Klaus Molmer, Phys. Rev. Lett. **68**, 580–583, 1992; and J. Optical Soc. Am. **B10**, 524, 1993.
29. R. Dum, P. Zoller and H. Ritsch, Phys. Rev. **A45**, 4879–4887, 1992; and C.W. Gardiner, A.S. Parkins and P. Zoller, Phys. Rev. **A46**, 4363, 1992.
30. S. Weinberg, "Dreams of a final theory", Hutchinson Radius, london, 1993.
31. N. Gisin, Phys. Lett. **143A**, 1–2 (1990).
32. D. Zeh, "Decoherence and Quantum measurements", Workshop on Stochastic Evolution of Quantum States in Open Systems and Measurements Processes, in press, Budapest, March 1993.
33. J.S. Bell, "Against "Measurement"", Physics World **3**, 33–40, 1990.
34. N. Gisin, Am. J. Phys. **61**, 86, 1993.

STOCHASTIC REPRESENTATION OF QUANTUM DYNAMICS

Z. HABA
Institute of Theoretical Physics
University of Wroclaw,
Wroclaw, Poland

1. Introduction

Since the appearance of the Feynman's paper [1] on the interpretation of quantum mechanics as a sum over classical paths there have been numerous attempts to make Feynman's sum over paths mathematically rigorous. An approach suggested by Gelfand and Yaglom [2] to use the propagation kernel in order to construct a complex measure appeared wrong [3]. Another approach based on a Fourier transform suggested first by Ito [4] and developed by Albeverio and Hoegh–Krohn [5] defines a complex measure (Fresnel integral), which however is not supported by paths in the configuration space. A relation of such a measure to summation over polygonal paths is discussed in Elworthy and Truman [6]. There have been numerous works concerned with an analytic continuation of the Wiener integral to the Feynman integral; let us mention Cameron [3], Ito [7], Nelson [8] (see also ref. [9]).

The difficulty with a naive definition of the Feynman integral originates from the non–existence of the Lebesgue measure in infinite number of dimensions. There are many reasons to use the Wiener measure in infinite number of dimensions to fulfill the role of the Lebesgue measure. An interpretation of to the Feynman integral as a generalized Wiener functional (i.e. a distribution) in infinite number of dimensions has been discussed in refs. [10]. The class of functionals which are integrable in various approaches is quite restricted. If we confine ourselves to analytic wave functions and severely restrict the class of potentials, which are admitted, then the Feynman integral can be expressed by the Wiener integral according to Cameron [3]. This expression has a rigorous mathematical meaning for a class of potentials discussed by Doss [11] and Azencott and Doss [12], who developed some ideas of Cameron [3].

We discuss here an extension of the approach of Cameron–Doss–Azencott. We extend the class of potentials which can be treated this way to Doss potentials [11] plus a Fourier transform of a bounded measure. We admit arbitrary (square integrable) wave functions expressing the Feynman propagator by the Wiener integral (the Feynman propagator is then used to define the time evolution). Next, we discuss a semi–classical expansion. We show that this expansion is asymptotic for sufficiently small time.

Z. Haba et al. (eds.), Stochasticity and Quantum Chaos, 73–90.
© 1995 *Kluwer Academic Publishers.*

2. Brownian motion representation of the solution of the Schroedinger equation

Let $\mathcal{F}_n$ denote the Banach algebra of functions of the form

$$f(\mathbf{x}) = \int exp(i\gamma\mathbf{x})d\nu_f(\gamma) \tag{2.1}$$

where ν is a complex measure on R^n with a bounded variation $|\nu|$ (this is the class of functions discussed in refs. [4]–[5]). We have to restrict this class further to analytic functions (such a restriction has been also introduced in ref. [13]). We need still another restriction, which comes out in one step of the proof of Theorem 2.2 below. So, we consider the set $\mathcal{H}_n^{\mathcal{F}} \subset \mathcal{F}_n$ defined by

$$\underset{\epsilon > 0}{\forall} \underset{K}{\exists} \underset{a \in R}{\forall} \quad \int exp(a|\gamma|) \, d|\nu|(\gamma) \leq K \, (\epsilon) exp(\epsilon a^2) \tag{2.2}$$

Both $\mathcal{F}_n$ and $\mathcal{H}_n^{\mathcal{F}}$ are dense subsets in $L^2(R^n)$.

We are interested in a path integral representation of the solution of the Schroedinger equation

$$i\hbar\partial_t\psi_t(\mathbf{x}) = \left(-\frac{\hbar^2}{2m}\Delta + gV(\mathbf{x})\right)\psi_t(\mathbf{x}) \equiv (H\psi_t)(\mathbf{x})$$

$$\psi\Big|_{t\,t=0} = \psi \tag{2.3}$$

where $\psi, V \in \mathcal{H}_n^{\mathcal{F}}$. We shall also consider a complex extension of eq. (2.3) in the form

$$\partial_t\psi_t(\mathbf{x}) = \left(\frac{\hbar\lambda^2}{2m}\Delta + \frac{g}{\hbar\lambda^2}V(\mathbf{x})\right)\psi_t \tag{2.4}$$

where $\lambda \in \mathbb{C}$ (clearly eq. (2.3) corresponds to $\lambda = \sqrt{i} = \frac{1}{\sqrt{2}}(1+i)$).

$$\sigma = \left(\frac{\hbar}{m}\right)^{\frac{1}{2}}$$

b_t is the Brownian motion i.e. the Gaussian process with the covariance

$$E\left[b^k(t)b^r(s)\right] = \delta_{kr}min(t,s) \qquad k, r = 1, 2, ..., n \tag{2.5}$$

We need first the following

Lemma 2.1
Let μ be a Gaussian measure on the Wiener space $C([0, t])$. Assume that Q is a non–negative quadratic form on $C([0, t])$ such that $Q = \underset{n \to \infty}{lim} Q_n$ with probability 1, where Q_n is a continuous non–negative quadratic form. Let $\mathcal{L}(b) = \sum_{k=1}^{m} a_k b(\tau_k)$,

where $a_k \in R$. Then,

$$\left| \int d\mu(b) exp\left(-Q(b) - i\lambda \mathcal{L}(b)\right) \right| \leq 1 \qquad (2.6)$$

if $\text{Re}(\lambda^2) \geq 0$.

Proof : for Q_n we have explicitly (see ref. [14], Theorem 4, sec. 18)

$$\int d\mu(b) exp\left(-Q_n(b) - i\lambda \mathcal{L}(b)\right) = det\left(1 + G^{\frac{1}{2}} Q_n G^{\frac{1}{2}}\right)^{-\frac{1}{2}}$$

$$exp\left(-\frac{\lambda^2}{2} \sum_{r,s=1}^{m} c_r c_s C_n(t_r, t_s)\right)$$

where

$$C_n = G^{\frac{1}{2}} \left(1 + G^{\frac{1}{2}} Q_n G^{\frac{1}{2}}\right)^{-1} G^{\frac{1}{2}}$$

where G is the covariance of μ. It follows that the bound (2.6) holds true for Q_n. Then, $exp(-Q_n - i\lambda\mathcal{L})$ is bounded by an integrable function $exp|\lambda\mathcal{L}|$. Hence, eq. (2.6) holds true by Lebesgues dominated convergence.

With a general potential of the form (2.2) we need a cutoff on an intermediate stage

$$\theta_R(\mathbf{x}) = exp\left(-\frac{|\mathbf{x}|^2}{2R}\right)$$

Let us denote

$$\Omega_t\left(g\theta_R(\mathbf{b}); V(\mathbf{x} + \lambda\sigma\mathbf{b})\right) = exp\left\{\frac{g}{\hbar\lambda^2} \int_0^t \theta_R(\mathbf{b}_\tau) V(\mathbf{x} + \lambda\sigma\mathbf{b}_\tau) d\tau\right\} \qquad (2.7)$$

We shall also use a shorthand notation $\Omega(g\theta_R)$ for the left hand side of eq. (2.7).

Theorem 2.2
Assume that $\psi, V \in \mathcal{H}_n{}^{\mathcal{F}}$. Then, $E\left[\Omega_t(g\theta_R)\right]$ is an analytic function of $g \in \mathbb{C}$ and $\lambda \neq 0$ as long as $Re\lambda^2 \geq 0$. Define

$$\psi_t{}^R(\mathbf{x}) = E\left[\Omega_t(g\theta_R)\theta_R(\mathbf{b}_t)\psi(\mathbf{x} + \lambda\sigma\mathbf{b}_t)\right] \qquad (2.8)$$

Then, $\psi_t(\mathbf{x}) \equiv \lim_{R\to\infty} \psi_t{}^R(\mathbf{x})$ exists uniformly in t and $\mathbf{x}$. $\psi_t(\mathbf{x})$ solves the Schroedinger equation (2.4). The solution is an analytic function of g and λ (in the same region where $E\left[\Omega_t\right]$ is analytic).

Proof : The functional under the expectation value in eq. (2.8) is bounded in b. In fact, the exponent in eq. (2.7) is bounded because

$$\left|\theta_R(\mathbf{b}_\tau)V(\mathbf{x} + \lambda\sigma\mathbf{b}_\tau)\right| = \left|\int d\nu_V(\gamma) exp\left(-\frac{\mathbf{b}_\tau{}^2}{2R}\right) exp\left(i\gamma(\mathbf{x} + \lambda\sigma\mathbf{b}_\tau)\right)\right| \leq$$

$$\leq \int d|\nu_V|(\gamma) \, exp\left(|\gamma| \, |\lambda\sigma\mathbf{b}_\tau| - \frac{\mathbf{b}_\tau^2}{2R}\right) \leq K(R) \qquad (2.9)$$

owing to the assumption (2.2).

Hence, Ω as well as $\theta\psi$ are bounded. We expand Ω in power series in g. Then, applying the Lebesgue dominated convergence theorem we exchange the sum with the expectation value. We get

$$\psi_t^R = \lim_{N\to\infty} F_N(R) \qquad (2.10)$$

where

$$F_N(R) = \sum_{n=0}^{N} \frac{1}{n!} \left(\frac{q}{\hbar\lambda^2}\right)^n E\left[\int_0^t d\tau_1 .... \int_0^t d\tau_n d\nu_V(\gamma_1).....d\nu_V(\gamma_n)d\nu_\psi(\beta)\right.$$

$$\left. exp\left(-\mathcal{B}_n^R + i\left(\sum_{k=1}^{n} \gamma_k + \beta\right)\mathbf{x}\right)\right]$$

with

$$\mathcal{B}_n^R = \frac{1}{2R}\left(\mathbf{b}_t^2 + \sum_{k=1}^{n} \mathbf{b}_{\tau_k}^2\right) - i\sigma\lambda\left(\beta \, \mathbf{b}_t + \sum_{k=1}^{n} \gamma_k \mathbf{b}_{\tau_k}\right)$$

Then, owing to the assumption (2.2) we can apply the Fubini theorem

$$E\left[\int d\nu_V(\gamma_1).....d\nu_V(\gamma_n)d\nu_\psi(\beta) \, exp\left(-\mathcal{B}_n^R\right)\right] =$$

$$= \int d\nu_V(\gamma_1)....d\nu_V(\gamma_n)d\nu_\psi(\beta) E\left[exp\left(-\mathcal{B}_n^R\right)\right]$$

Now, we are able to show the existence of the limit

$$\lim_{R\to\infty} F_N(R) \equiv F_N \qquad (2.11)$$

This follows from the Lebesgue dominated convergence theorem, because the measure $|\nu|$ is finite and

$$\left|E\left[exp(-\mathcal{B}_n^R)\right]\right| \leq 1 \qquad (2.12)$$

by virtue of Lemma 2.1. Note that from the bound (2.12) it follows that

$$\left|F_N(R) - F_M(R)\right| \leq \sum_{n=M}^{N} \left(\frac{|g|t}{\hbar|\lambda|^2}\right)^n \frac{|\nu|^n}{n!}$$

where $|\nu|$ is the total variation of ν.

Hence, the convergence $N \to \infty$ of $F_N(R)$ is uniform in R (as well as in $\mathbf{x}$ and t).

Therefore, we may exchange the limits $N \to \infty$ and $R \to \infty$

$$\psi_t \equiv \lim_{R \to \infty} \psi_t{}^R = \lim_{R \to \infty} \lim_{N \to \infty} F_N(R) = \lim_{N \to \infty} \lim_{R \to \infty} F_N(R) =$$

$$= \sum_n^{\infty} (\lambda^{-2})^n \, \hbar^{-n} \, \frac{g^n}{n!} E\left[\left(\int V_\tau(\mathbf{x} + \lambda\sigma\mathbf{b}_\tau) d\tau \right)^n \psi(\mathbf{x} + \lambda\sigma\mathbf{b}_t) \right] =$$

$$= \sum_n^{\infty} (\lambda^{-2})^n \, \hbar^{-n} \, \frac{g^n}{n!} \int_0^t d\tau_1 ... \int_0^t d\tau_n \int d\nu_V(\gamma_1) ...... d\nu_V(\gamma_n) d\nu_\psi(\beta) \qquad (2.13)$$

$$exp\left\{ \left(i\sum_k \gamma_k + i\beta \right) \mathbf{x} \right\} exp\left\{ -\tfrac{1}{2}\sigma^2\lambda^2 E\left[\left(\sum_k \gamma_k \mathbf{b}_{\tau_k} + \beta\mathbf{b}_t \right)^2 \right] \right\}$$

It is clear from eq. (2.13) that if $|\nu_V|$ and $|\nu_\psi|$ are bounded, $\lambda \neq 0$ and $Re\ \lambda^2 \geq 0$ then the series (2.13) is convergent uniformly in t and $\mathbf{x}$. It can also be checked by direct calculation that ψ_t (2.13) fulfills the Schroedinger equation.

As a further extension of the formula (2.8) we consider the Schroedinger equation in an electromagnetic field

$$\partial_t \psi_t(\mathbf{x}) = \left(\frac{\hbar\lambda^2}{2m} \nabla_A{}^2 + \frac{g}{\hbar\lambda^2} V_t \right) \psi_t \qquad (2.14)$$

where

$$(\nabla_A)_k = \partial_k + ieA_k$$

Denote

$$\Omega_t{}^A(g\theta_R) \equiv S_R{}^A \Omega(g\theta_R) = exp\left\{ \frac{i}{\hbar}\lambda\sigma e \int_0^t \theta_R(\mathbf{b}_\tau)\mathbf{A}(\mathbf{x} + \lambda\sigma\mathbf{b}_\tau) \circ d\mathbf{b}_\tau \right\}$$

$$exp\left\{ \frac{g}{\hbar\lambda^2} \int_0^t \theta_R(\mathbf{b}_\tau)V(\mathbf{x} + \lambda\sigma\mathbf{b}_\tau)d\tau \right\} \qquad (2.15)$$

Here, $\theta_R\mathbf{A}$ is a bounded function. The circle denotes the Stratonovitch stochastic integral (see ref. [17]).

Theorem 2.3
Assume $\mathbf{A}, V, \psi \in \mathcal{H}_n{}^{\mathcal{F}}$. Define

$$\psi_t{}^R(\mathbf{x}) = E\left[\Omega_t{}^A(g\theta_R) \psi(\mathbf{x} + \lambda\sigma\mathbf{b}_\tau) \right] \qquad (2.16)$$

Then, $\lim_{R \to \infty} \psi_t{}^R(\mathbf{x}) \equiv \psi_t(\mathbf{x})$ exists uniformly in t and $\mathbf{x}$. ψ_t is the unique solution of the Schroedinger equation (2.14) with the initial condition ψ. ψ_t is an analytic function of $g \in \mathbb{C}$, $e \in \mathbb{C}$ and $\lambda \neq 0$ as long as $Re\lambda^2 \geq 0$. $\psi_t(\mathbf{x})$ has an analytic continuation $\psi_t(\mathbf{x} + i\mathbf{y})$ to $\mathbb{C}^n$.

Proof : from eq. (2.15) $\Omega_t{}^A(g\theta_R) = S_R{}^A\Omega_t(g\theta_R)$ where the first factor is bounded by an integrable function S^A on the basis of the results of ref. [15] and the second factor is bounded by an integrable function on the basis of Theorem 2.2. So, Lebesgue dominated convergence theorem applies. Moreover, the estimates (2.20) and (2.12) imply that the power series in e and g is absolutely convergent (as in the proof of Theorem 2.2).

3. Representation of the Feynman propagator in terms of the Brownian bridge

We wish to extend the definition of U_t from $\mathcal{H}_n{}^{\mathcal{F}}$. This is a dense set in $L^2(R^n)$. So, a definition of U_t on this set determines U_t in a unique way. However, we cannot extend U_t to $L^2(R^n)$ directly through the formula (2.8). We start with a definition of a semigroup S_t on $L^2(R^{2n})$ and its kernel. Then, through a restriction of this kernel to analytic functions we are able to define the Feynman propagator $U_t(\mathbf{x}, \mathbf{y})$ for Doss potentials [11] as well as the potentials of the form of a Fourier transform of a bounded measure.

For an arbitrary complex–valued potential $\mathcal{V}$ on R^{2n} such that

$$Im\mathcal{V}(\mathbf{z}) \leq a|\mathbf{z}| \tag{3.1}$$

for a certain $a \geq 0$ we define a semigroup S_t on a set $\mathcal{G}$ consisting of functions such that

$$|\Phi(\mathbf{z})| \leq Kexp(c|\mathbf{z}|) \tag{3.2}$$

for certain $c \geq 0$ and $K \geq 0$. S_t is defined by the Feynman–Kac formula

$$(S_t\Phi)(\mathbf{z}) = E\left[exp\left(-\frac{i}{\hbar}\int_0^t \mathcal{V}\left(\mathbf{x}_1 + \frac{\sigma}{\sqrt{2}}\mathbf{b}_\tau, \mathbf{x}_2 + \frac{\sigma}{\sqrt{2}}\mathbf{b}_\tau\right)d\tau\right)\right.$$

$$\left.\Phi\left(\mathbf{x}_1 + \frac{\sigma}{\sqrt{2}}\mathbf{b}_t, \mathbf{x}_2 + \frac{\sigma}{\sqrt{2}}\mathbf{b}_t\right)\right] =$$

$$= \int E\left[\delta(\mathbf{y}_1 - \mathbf{x}_1 - \frac{\sigma}{\sqrt{2}}\mathbf{b}_t)\delta(\mathbf{y}_2 - \mathbf{x}_2 - \frac{\sigma}{\sqrt{2}}\mathbf{b}_t) \right. \tag{3.3}$$

$$exp\left(-\frac{i}{\hbar}\int_0^t \mathcal{V}(\mathbf{x}_1 + \frac{\sigma}{\sqrt{2}}\mathbf{b}_\tau, \mathbf{x}_2 + \frac{\sigma}{\sqrt{2}}\mathbf{b}_\tau)d\tau\right)\right]\Phi(\mathbf{y}_1, \mathbf{y}_2)d\mathbf{y}_1 d\mathbf{y}_2$$

$$\equiv \int S_t\left(\mathbf{x}_1, \mathbf{x}_2; \mathbf{y}_1, \mathbf{y}_2\right)\Phi\left(\mathbf{y}_1, \mathbf{y}_2\right)d\mathbf{y}_1 d\mathbf{y}_2$$

where

$$\sigma = \sqrt{\frac{\hbar}{m}}$$

In eq. (3.3) we used the decomposition of R^{2n} into $R^n \oplus R^n$ $z = (\mathbf{x}_1, \mathbf{x}_2)$ where $\mathbf{x}_1, \mathbf{x}_2 \in R^n$. The last two lines of eq. (3.3) define the kernel of the operator S_t.

We can express the kernel $S_t(\mathbf{x}_1, \mathbf{x}_2; \mathbf{y}_1, \mathbf{y}_2)$ of S_t by the Brownian bridge α. The Brownian bridge (see ref. [16]) is the Gaussian process defined on the interval [0,1] with the covariance

$$E\left[\alpha_k(s)\alpha_r(s')\right] = \delta_{kr} s(1-s') \quad \text{if} \quad s \le s' \tag{3.4}$$

and the boundary condition $\alpha(0) = \alpha(1) = 0$.

Then, a simple application of the arguments of ref. [16] leads to the formula

$$S_t(\mathbf{x}_1, \mathbf{x}_2; \mathbf{y}_1, \mathbf{y}_2) = (4\pi t\sigma^2)^{-\frac{n}{2}} \exp\left(-\frac{(\mathbf{x}_1 + \mathbf{x}_2 - \mathbf{y}_1 - \mathbf{y}_2)^2}{4t\sigma^2}\right) \delta(\mathbf{x}_1 - \mathbf{x}_2 - \mathbf{y}_1 + \mathbf{y}_2)$$

$$E\left[\exp\left\{-\frac{i}{\hbar}\int_0^t V\left(\mathbf{x}_1\left(1 - \frac{s}{t}\right) + \mathbf{y}_1\frac{s}{t} + \sqrt{\frac{t}{2}}\,\sigma\alpha\left(\frac{s}{t}\right),\right.\right.\right.$$

$$\left.\left.\left. \mathbf{x}_2\left(1 - \frac{s}{t}\right) + \mathbf{y}_2\frac{s}{t} + \sqrt{\frac{t}{2}}\,\sigma\alpha\left(\frac{s}{t}\right)\right)ds\right\}\right]$$

(3.5)

We restrict ourselves now to functions of the form

$$\mathcal{V}(\mathbf{x}_1, \mathbf{x}_2) \equiv V(\mathbf{x}_1 + i\mathbf{x}_2)$$

$$\Phi(\mathbf{x}_1, \mathbf{x}_2) \equiv \phi(\mathbf{x}_1 + i\mathbf{x}_2) \tag{3.6}$$

where V and ϕ are holomorphic functions of their arguments. Inserting eq. (3.5) into eq. (3.3) we get

$$(S_t\phi)(\mathbf{x}_1 + i\mathbf{x}_2) = (4\pi t\sigma^2)^{-\frac{n}{2}} \int d\mathbf{y}\,\phi\left(\frac{\mathbf{y} + \mathbf{x}_1 + \mathbf{x}_2}{2} + i\frac{\mathbf{y} - \mathbf{x}_1 + \mathbf{x}_2}{2}\right)$$

$$\exp\left(-\frac{(\mathbf{y} - \mathbf{x}_1 - \mathbf{x}_2)^2}{4t\sigma^2}\right) \mathcal{E}_t(\mathbf{y}, \mathbf{x}_1, \mathbf{x}_2) \tag{3.7}$$

where by $\mathcal{E}$ we denoted $E[.....]$ (the last factor in eq.(3.5)) with $\mathbf{y}_1 = \frac{\mathbf{y} + \mathbf{x}_1 - \mathbf{x}_2}{2}$ and $\mathbf{y}_2 = \frac{\mathbf{y} - \mathbf{x}_1 + \mathbf{x}_2}{2}$.

We show next that

$$(4\pi t\sigma^2)^{-\frac{n}{2}} \int d\mathbf{y}\,\chi\left(\frac{\mathbf{y} + \mathbf{x}}{2} + i\frac{\mathbf{y} - \mathbf{x}}{2}\right) \exp\left(-\frac{(\mathbf{x} - \mathbf{y})^2}{4t\sigma^2}\right) =$$

$$= (2\pi i t\sigma^2)^{-\frac{n}{2}} \int d\mathbf{y}\,\chi(\mathbf{y})\exp\left(-\frac{(\mathbf{x} - \mathbf{y})^2}{2it\sigma^2}\right) \tag{3.8}$$

(where $i^{\frac{n}{2}} = exp\frac{in\pi}{2}$) for any analytic function $\chi \in L^1(R^n)$ of the form

$$\chi(\mathbf{y}) = \int d\mathbf{p}\tilde{\chi}(\mathbf{p})exp\ i\mathbf{p}\mathbf{y} \tag{3.9}$$

In order to check eq. (3.8) we insert eq. (3.9) into eq. (3.8), apply the Fubini theorem and perform the y–integral.

Finally, we set $\mathbf{x}_2 = 0$ in eq. (3.7) and apply the identity (3.8). We get an expression for $(S_t\phi)(\mathbf{x})$ in terms of the kernel

$$K_t(\mathbf{x},\mathbf{y}) = (2\pi it\sigma^2)^{-\frac{n}{2}}exp\left(-\frac{(\mathbf{x}-\mathbf{y})^2}{2it\sigma^2}\right)$$

$$E\left[exp\left\{-\frac{i}{\hbar}\int_0^t V\left(\mathbf{x}(1-\frac{s}{t})+\mathbf{y}\frac{s}{t}+\lambda\sqrt{t}\sigma\alpha(\frac{s}{t})\right)ds\right\}\right] \tag{3.10}$$

$$\equiv K_t^0(\mathbf{x},\mathbf{y})\mathcal{R}_t(\mathbf{x},\mathbf{y})$$

where $\lambda = \sqrt{i} \equiv \frac{1}{\sqrt{2}}(1+i)$
and we denoted the free propagator by K^0.
Eq. (3.10) is our final formula for the Feynman propagator.

4. Definition of the Feynman kernel for potentials from $\mathcal{H}_n^{\mathcal{F}}$

As for the wave function (2.8) we first regularize the potential $V \rightarrow V^R$ in the kernel $\mathcal{R}$. We get this way the regularized kernels $\mathcal{R}^R_t$. We prove

Theorem 4.1
Define

$$\mathcal{R}_t^R(\mathbf{x},\mathbf{y}) = E\left[exp\left\{\frac{g}{\lambda^2\hbar}\int_0^t \theta_R(\alpha(\frac{s}{t}))V\left(\mathbf{x}(1-\frac{s}{t})+\mathbf{y}\frac{s}{t}+\lambda\sigma\sqrt{t}\alpha(\frac{s}{t})\right)ds\right\}\right]$$

$$\tag{4.1}$$

where $V \in \mathcal{H}_n^{\mathcal{F}}$. The limit $R \rightarrow \infty$ exists uniformly in $t,\mathbf{x},\mathbf{y}$. $\mathcal{R}$ is an analytic function of g and λ if $Re\lambda^2 \geq 0$.

Proof : We expand the exponential in eq. (4.1) in a power series in g. As in the proof of Theorem 2.2 using the Lebesgue dominated convergence theorem we exchange the sum with the expectation value. We get

$$\mathcal{R}_t^R = \lim_{N\rightarrow\infty} \mathcal{R}_N(R) \tag{4.2}$$

where

$$\mathcal{R}(R) = \sum_{n=0}^{N} \frac{1}{n!} \left(\frac{g}{\lambda^2 \hbar}\right)^n E\left[\int_0^t ds_1 .... \int_0^t ds_n dv_V(\gamma_1).....dv_V(\gamma_n)\right.$$

$$\left. exp\left(-\mathcal{B}_n{}^R + i\sum_{k=1}^{n} \gamma_k z(s_k)\right)\right]$$

with

$$\mathcal{B}_n{}^R = \frac{t}{2R}\sum_{k=1}^{n} \alpha\left(\frac{s_k}{t}\right)^2 - i\sigma\lambda\sqrt{t}\sum_{k=1}^{n} \gamma_k\alpha\left(\frac{s_k}{t}\right)$$

and

$$z(s) = x\left(1 - \frac{s}{t}\right) + y\frac{s}{t}$$

Again as in the proof of Theorem 2.2 the crucial argument for the uniform convergence

$$\lim_{R\to\infty} \mathcal{R}_N(R) \equiv \mathcal{R}_N \tag{4.3}$$

comes from the bound (by Lemma 2.1)

$$\left|E\left[exp\left(-\mathcal{B}_n{}^R\right)\right]\right| \le 1 \tag{4.4}$$

We can conclude that the convergence $N \to \infty$ of $\mathcal{R}_N(R)$ is uniform in R (as well as in x and t). Hence, we may exchange the limits $N \to \infty$ and $R \to \infty$

$$\mathcal{R} \equiv \lim_{R\to\infty} \mathcal{R}^R = \lim_{R\to\infty} \lim_{N\to\infty} \mathcal{R}_N(R) = \lim_{N\to\infty} \lim_{R\to\infty} \mathcal{R}_N(R) =$$

$$= \sum_n^{\infty}(\lambda^{-2})^n\hbar^{-n} \frac{g^n}{n!} E\left[\left\{\int V\left(z(\tau) + \lambda\sigma\sqrt{t}\alpha(\frac{\tau}{t})\right) d\tau\right\}^n\right)\right] =$$

$$= \sum_n^{\infty}(\lambda^{-2})^n\hbar^{-n} \frac{g^n}{n!} \int_0^t d\tau_1 ... \int_0^t d\tau_n \int dv_V(\gamma_1)......dv_V(\gamma_n)) \tag{4.5}$$

$$exp\left\{(i\sum_k \gamma_k z(\tau_k)\right\} exp\left\{-\frac{t}{2}\sigma^2\lambda^2 E\left[\left(\sum_k \gamma_k\alpha\left(\frac{\tau_k}{t}\right)\right)^2\right]\right\}$$

5. Semiclassical approximation to the solution of the Schroedinger equation

The semiclassical expansion of the wave function for the Doss class of potentials has been discussed by Azencott and Doss [12]. The expansion is based on the following lemma proved in ref. [12]

Lemma 5.1

Assume that $F(\mathbf{b} + \zeta\mathbf{f})$ is an analytic function of ζ. Then for any integrable F (depending on $\mathbf{b}(s)$ with $0 \le s \le t$) and any $\zeta \in \mathbb{C}$

$$E\left[F(\mathbf{b})\right] = E\left[F(\mathbf{b} + \zeta\mathbf{f})exp\left(-\frac{\zeta^2}{2}\int\limits_0^t (\frac{d\mathbf{f}}{ds})^2 ds\right) exp\left(-\zeta\int\limits_0^t \frac{d\mathbf{f}}{ds}d\mathbf{b}(s)\right)\right] \quad (5.1)$$

if $\mathbf{f}'$ is square integrable.

If $\zeta \in R$ then eq. (5.1) is the standard Cameron–Martin formula [17]. Then, because the right hand side is analytic in $\zeta \in \mathbb{C}$ and independent of ζ if $\zeta \in R$ it follows that it does not depend on $\zeta \in \mathbb{C}$.

Applying eq. (5.1) to ψ_t^R (2.8) we get the following identity

$$\psi_t^R(\mathbf{x}) = E\left[\Omega_t\left(g\theta_R(\mathbf{b} + \zeta\mathbf{f}); V(\mathbf{x} + \lambda\sigma\mathbf{b} + \lambda\sigma\zeta\mathbf{f})\right)\theta_R\left(\mathbf{b}_t + \zeta\mathbf{f}\right)\right.$$

$$\left.\psi\left(\mathbf{x} + \lambda\sigma\mathbf{b}_t + \zeta\lambda\sigma\mathbf{f}\right)exp\left(-\frac{\zeta^2}{2}\int\limits_0^t \left(\frac{d\mathbf{f}}{ds}\right)^2 - \zeta\int\limits_0^t \frac{d\mathbf{f}}{ds}d\mathbf{b}\right)\right] \quad (5.2)$$

We choose now $\zeta = \lambda^{-1}$. Let us denote

$$\boldsymbol{\xi}_s(\mathbf{x}) = \mathbf{x} + \sigma\mathbf{f}(s) \quad (5.3)$$

Assume that ψ has the semiclassical form

$$\psi = exp(\frac{iW}{\hbar})\phi \quad (5.4)$$

where

$$\phi(\mathbf{x}) = \int exp(i\mathbf{x}\boldsymbol{\gamma})d\nu_\phi(\boldsymbol{\gamma}) \quad (5.5)$$

We choose $\mathbf{f}$ in such a way that $\boldsymbol{\xi}$ is the solution of the boundary value problem

$$m\frac{d^2\boldsymbol{\xi}}{ds^2} = -\nabla V(\boldsymbol{\xi})$$

$$\boldsymbol{\xi}_s|_{s=0} = \mathbf{x} \quad \text{and} \quad m\frac{d\boldsymbol{\xi}}{ds}|_{s=t} = \nabla W(\boldsymbol{\xi}_t) \quad (5.6)$$

Next let us denote the classical action corresponding to the trajectory (5.6) by S

$$S_t(\boldsymbol{\xi}) = W(\mathbf{x}) + \frac{m}{2}\int\limits_0^t \left(\frac{d\boldsymbol{\xi}}{ds}\right)^2 ds - \int\limits_0^t gV(\boldsymbol{\xi}_s)ds \quad (5.7)$$

$S_t(\boldsymbol{\xi})$ is the solution of the classical Hamilton-Jacobi equation

$$\partial_t S + \frac{(\nabla S)^2}{2m} + V = 0 \quad (5.8)$$

with the initial condition $S_{t|t=0} = W$. The virtue of the boundary condition (5.6) is that the terms linear in **b** cancel in the formal limit $R \to \infty$. In this limit $\psi_t(\mathbf{x})$ depends on

$$V_2(\boldsymbol{\xi} + \lambda\sigma\mathbf{b}) \equiv V(\boldsymbol{\xi} + \lambda\sigma\mathbf{b}) - V(\boldsymbol{\xi}) - \nabla V(\boldsymbol{\xi})\lambda\sigma\mathbf{b} \tag{5.9}$$

which is of order $\hbar$. Hence,

$$\psi_t(\mathbf{x}) \sim exp\left(\frac{iS_t(\boldsymbol{\xi})}{\hbar}\right)\phi(\boldsymbol{\xi}_t) \tag{5.10}$$

We are going to give eq. (5.10) a rigorous meaning for $V \in \mathcal{H}_n{}^{\mathcal{F}}$ (for Doss potentials the semiclassical formula has been proven in ref. [12]).

Theorem 5.2
For t sufficiently small we have for each t and $\mathbf{x}$

$$\lim_{\hbar \to 0} \psi_t(\mathbf{x})exp(-\frac{i}{\hbar}S_t(\boldsymbol{\xi})) = \lim_{\hbar \to 0}\lim_{R \to \infty}\psi_t{}^R(\mathbf{x})exp(-\frac{iS_t(\boldsymbol{\xi})}{\hbar}) \equiv \Phi(t,\mathbf{x}) \tag{5.11}$$

If we choose $R = r\hbar^{-1}$ then the limit $r \to \infty$ is uniform in $\hbar$. Φ can be expressed explicitly by a convergent series

$$\Phi(t,\mathbf{x}) = \lim_{\hbar \to 0}\sum_n^\infty (\lambda^{-2})^n \frac{g^n}{n!}E\left[\left(\hbar^{-1}\int V_2(\boldsymbol{\xi}_\tau + \lambda\sigma\mathbf{b}_\tau)d\tau\right)^n \phi(\boldsymbol{\xi}_t + \lambda\sigma\mathbf{b}_t)\right] =$$

$$= \phi(\boldsymbol{\xi}_t)\sum_n^\infty (\lambda^{-2})^n(-2)^{-n}\frac{g^n}{n!}\int_0^t d\tau_1 ... \int_0^t d\tau_n \int d\nu_V(\boldsymbol{\gamma}_1)......d\nu_V(\boldsymbol{\gamma}_n)$$

$$\prod_k E\left[(\mathbf{b}_{\tau_k}\boldsymbol{\gamma}_k)^2\right]exp\left(i\sum_k \boldsymbol{\gamma}_k\boldsymbol{\xi}_{\tau_k}\right)$$

$$= E\left[exp\left\{\frac{1}{2m}\int_0^t \mathbf{b}(s)V''(\boldsymbol{\xi}_s)\mathbf{b}(s)ds\right\}\right]\phi(\boldsymbol{\xi}_t)$$

$$\tag{5.12}$$

where V'' is the second Frechet derivative.
Sketch of the proof: As in the proof of Theorem 2.2 we expand the expression under the expectation value in eq. (5.2) in powers of V_2 (5.9). In such a case it is easy to see that each term in the series is regular in $\hbar$. The proof of the convergence of the series involve a simple counting of the number of terms. The details of the proof will appear elsewhere.

6. Semiclassical expansion of the Feynman propagator

Let us first consider a simple case of a meromorphic potential V of the form

$$V(\mathbf{x}) = Q(\mathbf{x})P(\mathbf{x})^{-1} \tag{6.1}$$

where Q and P are polynomials and their degrees fulfill the inequality $deg(Q) \leq deg(P) + 2$. Then V is quadratically bounded except of a discrete set of points $\mathcal{D} \subset R^n$. The potential in eq. (3.10) can be singular at a discrete set of values of $\alpha(s)$. With such singularities the expectation value in eq. (3.10) still can be finite (such problems have been discussed in ref. [18]). However, in order to avoid eventual difficulties we restrict ourselves here to $\mathbf{x}$ and $\mathbf{y}$ from a set $R_0{}^n \equiv R^n - \mathcal{D}$ in the definition (3.10) of $K_t(\mathbf{x}, \mathbf{y})$. In order to determine $\mathcal{D}$ assume that $\mathbf{z}_0$ is the (complex) zero of $P(\mathbf{x})$. Then, the condition that the Brownian bridge hits the zero is

$$\mathbf{x}(1 - \frac{s}{t}) + \mathbf{y}\frac{s}{t} = Re\mathbf{z}_0 - Im\mathbf{z}_0 \tag{6.2}$$

for a certain $0 \leq s \leq t$ i.e. the zero should be on the line joining $\mathbf{x}$ and $\mathbf{y}$.

In order to derive a semiclassical expansion we make a shift of variables in eq. (3.10) similarly as we did it in eq. (5.2)

$$\alpha \to \alpha + \zeta\mathbf{f} \tag{6.3}$$

Now, we get

$$\mathcal{R}_t(\mathbf{x}, \mathbf{y}) = \mathcal{R}_t{}^\zeta(\mathbf{x}, \mathbf{y}) = exp\left(-\frac{\zeta^2}{2}\int\limits_0^1 (\frac{d\mathbf{f}}{d\tau})^2 d\tau\right)$$

$$E\left[exp\left\{-\zeta\int\limits_0^1 \frac{d\mathbf{f}(\tau)}{d\tau}d\alpha(\tau) - \frac{i}{\hbar}\int\limits_0^t V\left(\mathbf{x}(1 - \frac{s}{t}) + \mathbf{y}\frac{s}{t} + \lambda\sigma\sqrt{t}(\alpha(\frac{s}{t}) + \zeta\mathbf{f}(\frac{s}{t}))\right) ds\right\}\right] \tag{6.4}$$

It holds true for an arbitrary $\mathbf{f}$ defined on the interval $[0,1]$ whose derivative is square integrable and $\mathbf{f}(1) = \mathbf{f}(0) = 0$.

The potential V as well as the expectation value (6.4) is analytic in ζ. In such a case from the fact that $\mathcal{R}_t{}^\zeta(\mathbf{x}, \mathbf{y})$ in eq. (6.4) does not depend on ζ for ζ real we can conclude that it does not depend on ζ if ζ is allowed to be complex. We choose now $\zeta = \lambda^{-1}$ in eq. (6.4). Let us denote

$$\mathbf{q}(s; \mathbf{x}, \mathbf{y}) = \mathbf{x}\left(1 - \frac{s}{t}\right) + \mathbf{y}\frac{s}{t} + \sigma\sqrt{t}\,\mathbf{f}\left(\frac{s}{t}\right) \tag{6.5}$$

$\mathbf{q} \in R^n$ is a curve joining $\mathbf{x}$ with $\mathbf{y}$. We choose it as a solution of the Newton equation

$$\frac{d^2\mathbf{q}}{dt^2} = -\nabla V(\mathbf{q}) \tag{6.6}$$

with the boundary condition $\mathbf{q}(0) = \mathbf{x}$ and $\mathbf{q}(t) = \mathbf{y}$ (under our assumptions on V there exists a unique solution of eq. (6.6) at least for small time). We assume that $\mathbf{q} \neq Re\mathbf{z}_0 - Im\mathbf{z}_0$. Let us introduce the classical action

$$S_t(\mathbf{q}) = \frac{m}{2}\int\limits_0^t \left(\frac{d\mathbf{q}}{dt}\right)^2 ds - \int\limits_0^t V(\mathbf{q}(s))ds \tag{6.7}$$

Then, the formula for $\mathcal{R}$ (6.4) takes the form

$$\mathcal{R}_t(\mathbf{x},\mathbf{y}) = exp\frac{iS_t(\mathbf{q})}{\hbar} E\left[exp\left\{-\frac{i}{\hbar}\int\limits_0^t V_2\left(\mathbf{q}(s)+\lambda\sigma\sqrt{t}\,\alpha(\tfrac{s}{t})\right)ds\right\}\right]exp\left(\frac{(\mathbf{x}-\mathbf{y})^2}{2it\sigma}\right)$$

$$\equiv \mathcal{R}_t^{\hbar}(\mathbf{x},\mathbf{y})\,exp\frac{iS_t(\mathbf{q})}{\hbar}\,exp\left(\frac{(\mathbf{x}-\mathbf{y})^2}{2it\sigma}\right)$$

(6.8)

where

$$V_2(\mathbf{q}+\lambda\sigma\sqrt{t}\alpha)\equiv V(\mathbf{q}+\lambda\sigma\sqrt{t}\alpha)-V(\mathbf{q})-\lambda\sigma\sqrt{t}\alpha\nabla V(\mathbf{q})\qquad (6.9)$$

Theorem 6.1
Assume that $\mathbf{q}(s)\neq Rez_0 - Imz_0$ for any s and V is the rational function (6.1). Then for sufficiently small t

$$\lim_{\hbar\to 0} K_t(\mathbf{x},\mathbf{y})\left(2\pi it\sigma^2\right)^{\frac{n}{2}}\,exp\left(-\frac{iS_t(\mathbf{q})}{\hbar}\right) =$$

$$= E\left[exp\left\{\frac{t}{m}\int\limits_0^t\alpha(\tfrac{s}{t})V''(\mathbf{q}(s))\alpha(\tfrac{s}{t})ds\right\}\right]$$

(6.10)

where V'' denotes the second order Frechet derivative.

Moreover, the expansion in σ is asymptotic.

Proof : In order to prove that the expansion of $\mathcal{R}^{\hbar}$ in σ is asymptotic it is sufficient to show that $\mathcal{R}^{\hbar}$ and its all derivatives over σ are bounded uniformly in $0\leq\sigma\leq\epsilon$ for a certain ϵ. First, if V is a rational function (6.1) then V_2 (6.9) is also a rational function of the same type (quadratically bounded) with the same discrete set $\mathcal{D}$ of singularities. Now for Gaussian integrals we have [16]

$$E\left[\mathcal{P}(\alpha)exp\left(\frac{t}{2}\int\limits_0^1\omega(\tau)^2\alpha(\tau)^2 d\tau\right)\right] < \infty \qquad (6.11)$$

if $t^2 < \pi^2(sup_\tau |\omega(\tau)|^2)^{-1}$ for any polynomially bounded function $\mathcal{P}$ (this is a sufficient condition for a finiteness of the integral (6.11) not a necessary one). From eq. (6.11) it follows that $\mathcal{R}^{\hbar}$ is integrable uniformly in σ, because V_2 is quadratically bounded. Applying the Lebesgue dominated convergence we get the explicit formula (6.10) for the limit $\sigma\to 0$. It follows also from eq. (6.11) that the integral (6.10) is finite if

$$t^2\,\sup_s\,V''(\mathbf{q}(s)) < m\pi^2 \qquad (6.12)$$

There remains to show that the derivatives are also uniformly bounded. We have

$$\left| \frac{d}{d\sigma} (\mathcal{R}^{\hbar}{}_t) \right| \leq$$

$$E\left[\left| \frac{d}{d\sigma} \frac{1}{m\sigma^2} \int_0^t V_2(\mathbf{q}(\tau) + \lambda\sigma\sqrt{t}\alpha(\frac{\tau}{t}))d\tau \right| \right. \tag{6.13}$$

$$\left. exp\left\{ \frac{1}{m\sigma^2} \int_0^t ImV_2\left(\mathbf{q}(s) + \lambda\sigma\sqrt{t}\,\alpha(\frac{s}{t}) \right) \right\} \right]$$

For the rational potential (6.1) $\frac{d}{d\sigma}\sigma^{-2}V_2$ in eq. (6.13) is bounded uniformly in σ and is polynomially bounded in α. ImV_2 in the exponential in eq. (6.11) is bounded uniformly in σ by a quadratic form in α of the type (6.10). Hence, from the formula (6.11) it follows that the right hand side of eq. (6.13) is bounded uniformly in σ if time is small enough. It is clear from the argument applied to the first derivative that we can continue differentiation in σ and that the derivatives will be bounded uniformly in σ. Hence, the semiclassical expansion is asymptotic for sufficiently small time.

We would like to prove the semiclassical formula also for potentials of the form of a Fourier transform of a bounded measure. We get

Theorem 6.2
Assume $V \in \mathcal{H}_n{}^{\mathcal{F}}$ then

$$\lim_{\hbar \to 0} K_t(\mathbf{x}, \mathbf{y}) exp\left(-\frac{iS_t(\mathbf{q})}{\hbar} \right) (2\pi it\sigma^2)^{n/2} =$$

$$\lim_{\hbar \to \infty} \lim_{R \to \infty} K_t{}^R(\mathbf{x}, \mathbf{y}) exp\left(-\frac{iS_t(\mathbf{q})}{\hbar} \right) (2\pi it\sigma^2)^{n/2} = \tag{6.14}$$

$$E\left[exp\left(t\int_0^t \alpha\left(\frac{s}{t}\right) V''(\mathbf{q}(s))\alpha\left(\frac{s}{t}\right) ds \right) \right]$$

Moreover, the expansion in σ is asymptotic.
Sketch of the proof: after the shift (6.4)

$$\mathcal{R}_t{}^R(\mathbf{x}, \mathbf{y}) = E\left[exp\left(\frac{i}{2}\int_0^t (\frac{d\mathbf{f}}{ds})^2 - \lambda^{-1}\int_0^t \frac{d\mathbf{f}}{ds} d\alpha \right) \right.$$

$$\left. exp\left\{ \frac{g}{\lambda^2\hbar} \int_0^t \theta_R(\alpha(\frac{s}{t}) + \lambda^{-1}\mathbf{f}(\frac{s}{t}))V\left(\mathbf{q}(\frac{s}{t}) + \lambda\sigma\sqrt{t}\alpha(\frac{s}{t}) \right) ds \right\} \right] \tag{6.15}$$

We write

$$\theta_R(\alpha + \frac{\mathbf{f}}{\lambda})V\left(\mathbf{q} + \lambda\sigma\sqrt{t}\alpha\right) =$$

$$= \theta_R(\alpha + \frac{\mathbf{f}}{\lambda})\left(V(\mathbf{q}) + \lambda\sigma\sqrt{t}\alpha\nabla V(\mathbf{q}) + V_2(\mathbf{q} + \lambda\sqrt{t}\sigma\alpha)\right)$$

(6.16)

Now, as in the proof of Theorem 5.2 we expand in powers of V_2. V_2 is regular in $\hbar$. The main technical difficulty (as in the proof of Theorem 5.2) is the proof of the convergence of the series, which again involves an estimate on the number of terms resulting from the expansion (roughly speaking the series is convergent because V_2 is quadratically bounded; however, this can be seen only after the computation of expectation values). The details of the proof will appear elsewhere.

Let us now return to the rational potentials discussed at the beginning of this section. We are interested what happens for large time. If $deg P + 1 \geq deg Q$ then the formula (6.4) is well defined for arbitrarily large t. However, the remaining non-classical terms could give a large contribution for small σ. In order to get some feeling what can happen let us apply the Jensen inequality in the form

$$E\left[\int_0^t \frac{ds}{t}exp\left(A(s)\right)\right] \geq E\left[exp\left(\int_0^t \frac{ds}{t}A(s)\right)\right]$$

(6.17)

true for any real function A. Applying the Jensen inequality (6.17) and the Fubini theorem we can get the estimate

$$\left|\mathcal{R}_t(\mathbf{x},\mathbf{y})\right| \leq \int_0^t \frac{ds}{t}E\left[exp\left\{\frac{t}{\hbar}Im V_2\left(\mathbf{q}(s) + \lambda\sigma\sqrt{t}\alpha(\frac{s}{t})\right)\right\}\right] =$$

$$\int_0^t \frac{ds}{t}\int d\mathbf{z}(2\pi s)^{-\frac{n}{2}}(t-s)^{\frac{n}{2}}exp\left(-\frac{(t-s)\mathbf{z}^2}{2s}\right)$$

(6.18)

$$exp\left\{\frac{t}{\hbar}Im V_2\left(\mathbf{q}(s) + \lambda\sigma\sqrt{t}\left(1-\frac{s}{t}\right)\mathbf{z}\right)\right\}$$

In order to compute the expectation value (6.18) we used the representation of the Brownian bridge in terms of the Brownian motion $\alpha(\tau) = (1-\tau)\mathbf{b}(\frac{\tau}{1-\tau})$ (see ref. [16]).
$Im V_2$ has the expansion in powers of σ

$$\frac{t}{\hbar}Im V_2 = a_2\mathbf{y}^2 + \sigma a_3\mathbf{y}^3 + ...$$

In spite of this regularity in σ the integral (6.18) in general behaves as $exp(\frac{K}{\hbar})$, where the constant K can be positive. If this is the true behavior of the functional integral (6.8) then we have a profound departure from the semiclassical approximation.

If in eq. (6.1) $deg(P) \geq deg(Q) - 1$ then $|\mathcal{R}_t(\mathbf{x},\mathbf{y})|$ (6.8) is bounded for arbitrarily large time t. However, for large time we are unable to get a bound uniform in $\hbar$.

From eq. (6.18) we can easily get a bound for arbitrary t of the form $\exp(\frac{K(t)}{\hbar})$, where $K(t)$ is positive. Apparently, the estimate $exp\frac{K}{\hbar}$ is too pessimistic (a blow–up of the semiclassical approximation). Deriving the bound (6.18) we took the absolute value under the integral sign neglecting the oscillatory terms in eq. (6.8) which seem to be crucial for the sign of K. Nevertheless, the formula (6.8) gives a sound starting point for an investigation of the large t and small $\hbar$ behavior. One should apply to the functional integral the saddle point method.

7. Quantum dynamics as stochastic classical mechanics

The Schroedinger equation (with an electromagnetic field)

$$i\hbar\partial_t\psi = \frac{1}{2m}\left(i\hbar\partial_\mu + eA_\mu\right)^2\psi + V\psi \tag{7.1}$$

(where $\mathbf{A}$ is a magnetic vector potential, V is a scalar potential and e is an electric charge) does not show any similarity with the classical mechanics defined as a flow on the phase space.
However, let us we make a similarity transformation

$$\psi_t = \chi_t\phi_t \tag{7.2}$$

where

$$\chi_t(\mathbf{x}) = exp\left(\frac{i}{\hbar}W_t(\mathbf{x})\right) \tag{7.3}$$

Then, ψ_t fulfills eq. (7.1) if and only if ϕ_t fulfills the equation

$$\partial_t\phi_t(\mathbf{x}) = \left(\frac{i\hbar}{2m}\Delta - \frac{1}{m}(\nabla W - e\mathbf{A})\nabla\right)\phi_t(\mathbf{x}) - \frac{1}{2m}div(\nabla W - e\mathbf{A})\phi_t(\mathbf{x}) -$$
$$- \frac{i}{\hbar}\left(\partial_t W + \frac{1}{2m}(\nabla W - e\mathbf{A})^2 + V\right)\phi_t(\mathbf{x}) \tag{7.4}$$

We get particularly simple form of eq. (7.4) if χ_t is any solution of the Schroedinger equation (7.1) (it does not need to be square integrable). Then, eq. (7.4) reads

$$\partial_t\phi = \left(\frac{i\hbar}{2m}\Delta - \frac{1}{m}(\nabla W - e\mathbf{A})\nabla\right)\phi \tag{7.5}$$

Note that in the formal limit $\hbar \to 0$ we get in eq. (7.5) the flow generated by

$$\frac{1}{m}(\nabla W - e\mathbf{A})\nabla$$

which is determined by the solution of the equation

$$\frac{d\mathbf{x}}{dt} = -\frac{1}{m}(\nabla W_t - e\mathbf{A}) \tag{7.6}$$

This is the classical flow

$$(U_t \phi)(\mathbf{x}) = exp \left(-\frac{1}{2m} \int\limits_0^t div(\nabla W_s - e\mathbf{A})(q_s)ds \right) \phi(q_t(\mathbf{x})) \qquad (7.7)$$

where $q_t(\mathbf{x})$ is the solution of eq. (7.6) with the initial condition $\mathbf{x}$.

The relation with the classical mechanics is visible if we express the Schroedinger equation for χ in terms of W

$$\partial_t W + \frac{(\nabla W - e\mathbf{A})^2}{2m} + V - \frac{i\hbar}{2m}(\triangle W - ediv\mathbf{A}) = 0 \qquad (7.8)$$

So, W in a (formal) limit $\hbar \to 0$ fulfills the Hamilton-Jacobi equation. If W is a general solution of the Hamilton–Jacobi equation (depending on n parameters) then eq. (7.6) is the solution of the Newton equation with the initial momenta determined by the above mentioned parameters (otherwise the solutions run over a Lagrangian submanifold of the phase space).

In order to get a classical (stochastic) representation of the solution of the second order equation (7.5) we introduce a complex diffusion process $\mathbf{q}_t(\mathbf{x})$ as a solution of the stochastic differential equation

$$d\mathbf{q}_t = -\frac{1}{m}(\nabla W - e\mathbf{A})dt + \lambda\sigma d\mathbf{b}_t \qquad (7.9)$$

with the initial condition $\mathbf{q}_{|t=0} = \mathbf{x}$ (where $\lambda = \frac{1}{\sqrt{2}}(1 + i)$).

Assume that there exists the unique solution $\mathbf{q}_t(\mathbf{x})$ of eq. (7.9) till the explosion time $\tau(\mathbf{q})$. Then

$$\psi_t(\mathbf{x}) = \chi_t(\mathbf{x})E[\phi(\mathbf{q}_t(\mathbf{x}))] \qquad (7.10)$$

is the unique solution of the Schroedinger equation (7.1) with the initial condition $\psi_0 = \chi_0\phi$.

It is important for applications to establish a relation between the Feynman formula (2.8) and the stochastic equation (7.9). In some cases it is easier to solve the stochastic equation (7.9). Then, as we have shown the formula (7.10) gives a solution of the Schroedinger equation(see ref. [19] for more details). On the other hand stochastic differential equations of the type (7.9) are of interest for themselves as models of various physical processes. Then, the Feynman integral may be useful for a study of their behaviour. It should be pointed out that to the complex equations (7.9) the methods of Lyapunov do not apply, because a positive definite Lyapunov function does not exist.

In order to relate eq. (7.9) to the Feynman integral it is sufficient to apply the formula (5.1) for a shift of variables with a special choice of $\mathbf{f}$. So, let us choose $\mathbf{f} = -\frac{1}{m\sigma}(\nabla W - e\mathbf{A})(\mathbf{q})$ where $\mathbf{q}$ is the solution of eq. (7.9). In such a case the exponential factor in eq. ((5.1) is

$$exp \left(\frac{i}{2m\hbar} \int\limits_0^t (\nabla W - e\mathbf{A})^2 + i\frac{1}{m\sigma}\lambda \int\limits_0^t (\nabla W - e\mathbf{A})db \right) \qquad (7.11)$$

Next, we apply Ito differentiation formula (see ref. [17] ; keep in mind that W depends on $\mathbf{x} + \lambda\sigma b$)

$$dW = \partial_s W ds + \lambda\sigma\nabla W d\mathbf{b} + \frac{\lambda^2\sigma^2}{2}\,\triangle W ds \qquad (7.12)$$

We rewrite eq. (7.12) in an integral form, then using this form of eq. (7.12) we replace the stochastic integral $\int_0^t \nabla W d\mathbf{b}$ in eq. (7.11) by an ordinary integral. Finally, we use the Hamilton–Jacobi equation (7.8) relating W and V. Then eq. (7.11) takes the form

$$exp\left(-\frac{i}{\hbar}\int_0^t V\left(\mathbf{x} + \lambda\sigma b_s\right)ds + \frac{iW_t}{\hbar} - \frac{iW}{\hbar} + \frac{i\lambda\sigma}{\hbar}\int_0^t \mathbf{A}\circ d\mathbf{b}\right) \qquad (7.13)$$

which is the Feynman formula with an electromagnetic field (the circle denotes the Stratonovitch integral [17]).

This way we have shown both analytically (eqs. (7.5) and (7.9)) and through a transformation on paths that calculating the Feynman integral is equivalent to solving the stochastic differential equation (7.9) (see ref. [20] ; Euclidean version has been discussed earlier in ref. [21]).

References

1. R.P. Feynman, Rev. Mod. Phys. **20**, 367(1948)
2. I.M. Gelfand and A.M. Yaglom, Journ. Math. Phys. **1**, 48(1960)
3. R.H. Cameron, J. Math. and Phys. **39**, 126(1960)
4. K. Ito, Proceed. Fifth Berkeley Symp., Univ. Calif. Press, J. Neyman, Ed., 1967
5. S. Albeverio and R. Hoegh–Krohn, Mathematical Theory of Feynman Path Integrals, Springer, 1976
6. K.D. Elworthy and A. Truman, Ann. Inst. H. Poincare, A**41**, 115(1984)
7. K. Ito, Proceed. Fourth Berkeley Symp., Univ. Calif. Press, J. Neyman, Ed., 1961
8. E. Nelson, Journ. Math. Phys. **5**, 332(1964)
9. J. Feldman, Trans. Am. Math. Soc. **10**, 251(1963)
10. L. Streit and T. Hida, Stoch. Proc. Appl. **16**, 55(1983)
    M. de Faria, J. Potthoff and L. Streit, Journ. Math. Phys.**32**, 2123(1991)
11. H. Doss, Commun. Math. Phys. **73**, 247(1980)
12. R. Azencott and H. Doss, in Lect. Notes in Math. No. 1109, Springer, 1985
13. S. Albeverio and R. Hoegh–Krohn, Inv. Math. **40**, 59(1977)
14. A.V. Skorohod, Integration in Hilbert Spaces, Springer, 1974
15. E. Carlen and P. Kree, Annals Prob. **19**, 354(1991)
16. B. Simon, Functional Integration and Quantum Physics, Academic, 1979
17. N. Ikeda and S. Watanabe, Stochastic Differential Equations and Diffusion Processes, North Holland, 1981
18. H. Ezawa, J.R. Klauder, L.A. Shepp J. Math. Phys. **16**, 783(1975)
19. Z. Haba, Phys. Lett. **175A**, 371(1993)
20. Z. Haba, preprint IFT UWR No. 828, March 1993
21. S. Albeverio, K. Yasue and J.C. Zambrini, Ann. Inst. H. Poincare, A**49**, 259, (1989)
    A.B. Cruzeiro and J.C. Zambrini, J. Funct. Anal. **96**, 62(1991)

LEVEL REPULSION AND EXCEPTIONAL POINTS

W.D. HEISS[†] AND W.-H. STEEB[‡]
[†] *Centre for Nonlinear Studies and Department of Physics*
University of the Witwatersrand
PO Wits 2050, Johannesburg, South Africa
[‡] *Department of Applied Mathematics and Nonlinear Studies*
Rand Afrikaans University, PO Box 524
Johannesburg 2000, South Africa

Abstract. The Riemann sheet structure of the energy levels $E_n(\lambda)$ of an N-dimensional symmetric matrix problem of the form $H_0 + \lambda H_1$ is discussed. It is shown that the singularities of the energy levels in the complex λ-plane are related to avoided level crossings. In the vicinity of a complex conjugate pair of exceptional points the full matrix problem behaves locally like a two dimensional problem. The significance of the exceptional points for the non-linear system of differential equations which determine the energy levels and state vectors as functions of λ is discussed.

1. Introduction

Avoided energy level crossing occurs in virtually all branches of physics. It has been discussed in the early days of quantum mechanics[1] and has again attracted interest more recently in the context of discussions about quantum chaos.[2] An aspect that is not so widely known is the correlation between the occurrence of avoided level crossing and related singularities of the energy levels in the complex plane.

In this paper we address the connection between level crossing and exceptional points[3] of the Hamilton operator. This connection is of a similar nature to the connection between the poles of a scattering function and the resonance structure of a cross section. In the same way as the poles of the scattering function give rise to the shape of a cross section, the exceptional points bring about the shape of the spectrum, particularly the occurrence of avoided level crossing.

The comparison between the poles of a scattering function and exceptional points is to be taken with care as exceptional points are square root branch points and thus produce a Riemann sheet structure. Yet there are parallels in that singularities which lie close to each other can have a smoothing or cancelling effect upon the physical quantities like the cross section or, in our case, the energy levels.

In this contribution we first discuss the general Riemann sheet structure of the eigenvalues $E_n(\lambda)$ of an N-dimensional symmetric matrix eigenvalue problem of the form $H_0 + \lambda H_1$. As the general structure is somewhat complicated the cases $N = 2$ and $N = 3$ are more explicitly dealt with for illumination. Further it is shown that an isolated complex conjugate pair of singularities is directly related to avoided level crossing of two levels. This situation corresponds locally to the coupling of only two levels, i.e. to a 2×2 symmetric matrix problem. We further show that when two complex conjugate pairs of singularities lie close to each other, the avoided level

91

Z. Haba et al. (eds.), Stochasticity and Quantum Chaos, 91–98.

crossing may be affected in that its occurrence is either shifted to another position or cancelled altogether. In this case the situation is usually locally equivalent to a 3×3 symmetric matrix problem.

In the following section we define the problem and briefly recapitulate the definition and significance of exceptional points. Section three presents examples. In section four it is shown that two adjacent energy levels of an $N \times N$ matrix problem behave locally like the levels of a two dimensional matrix problem. In the last section we discuss the significance of the exceptional points in the set of non-linear differential equations by which the energy levels $E_n(\lambda)$ are determined.

2. Exceptional points

The eigenvalues $E_n(\lambda)$ of the symmetric $N \times N$ matrix $H_\lambda = H_0 + \lambda H_1$, λ real, are the roots of the algebraic equation

$$\det(E - H_\lambda) = 0. \tag{1}$$

We assume that no degeneracy of the eigenvalues occurs for all real values of λ. However, when the eigenvalues are continued into the complex λ-plane, there are specific complex λ-values where the eigenvalues coalesce.[4-6] These values are the exceptional points of H_λ. To enforce coalescence of the roots of Eq.(1) the additional algebraic equation

$$\frac{\mathrm{d}}{\mathrm{d}E} \det(E - H_\lambda) = 0 \tag{2}$$

must be obeyed simultaneously. Note that Eq.(1) is a polynomial of order N in E and λ, while Eq.(2) is a polynomial of order $N - 1$ in the two variables. By the theory of elimination[7] we eliminate from Eqs.(1) and (2) the variable E to obtain what is called the resultant[7] of Eq.(1) which is a polynomial in λ of order $N(N-1)$. Since the coefficients of this polynomial are real, the zeros occur in $N(N-1)/2$ complex conjugate pairs.

By construction the zeros of the resultant must coincide with the zeros of the discriminant[7] of Eq.(1) which is given by

$$D(\lambda) = \prod_{m<n} \Big(E_m(\lambda) - E_n(\lambda)\Big)^2. \tag{3}$$

In other words, with suitable choice of a constant factor A_N the resultant and the discriminant must be equal which means that we can write

$$D(\lambda) = A_N \prod_{m<n} (\lambda - \lambda_{mn})(\lambda - \bar{\lambda}_{mn}) \tag{4}$$

where the zeros λ_{mn} and their complex conjugate values $\bar{\lambda}_{mn}$ of the resultant are arranged so that when $\lambda = \lambda_{mn}$ or $\lambda = \bar{\lambda}_{mn}$, then only the term $(E_m - E_n)^2$ vanishes in Eq.(3).

It now follows that the function

$$F(\lambda) = \sqrt{D(\lambda)} = \prod_{m<n} (E_m(\lambda) - E_n(\lambda)) \tag{5}$$

has square root branch points only at the roots of the resultant which are the simultaneous roots of Eqs.(1) and (2). They are the exceptional points of H_λ. If λ circles the point λ_{mn} only the term $(E_m - E_n)$ changes sign in $F(\lambda)$. Consequently, if $E_m(\lambda)$ and $E_n(\lambda)$ coalesce at the complex value λ_{mn} they must be connected by a square root branch point, which means that the expansions

$$E_m(\lambda) = E_{mn} + \sum_{k=1}^{\infty} a_k (\sqrt{\lambda - \lambda_{mn}})^k$$

$$E_n(\lambda) = E_{mn} + \sum_{k=1}^{\infty} a_k (-\sqrt{\lambda - \lambda_{mn}})^k$$

(6)

exist. The radius of convergence is determined by the nearest next singularity where E_m or E_n is connected to a further of the N eigenvalues. Since all possible pairs E_m and E_n are connected in this way, the global Riemann structure is such that there is a global function $E(\lambda)$ which has N Riemann sheets. While this function is not explicitly known (for $N \geq 5$), the values assumed on the N sheets for real λ are just the eigenvalues $E_n(\lambda)$ of H_λ. The sheets are connected by $N(N-1)$ square root branch points which occur in complex conjugate pairs where any two of the N eigenvalues coalesce.

In the following section we illuminate these findings for low values of N.

3. Examples

The characteristic equation Eq.(1) is always of the form

$$\sum_{k=0}^{N} c_k(\lambda) E^{N-k} = 0$$

(7)

where $c_k(\lambda)$ is a polynomial in λ of order k with $c_0 = 1$. For $N = 2$ the solution of Eq.(7) reads

$$E_{1,2}(\lambda) = -\frac{c_1}{2} \pm \frac{1}{2}\sqrt{c_1^2 - 4c_2}.$$

(8)

Here we recognise the discriminant $c_1^2 - 4c_2$ which is a polynomial of second order in λ. Note that it is non-negative for real λ as E_1 and E_2 are real. Its zeros are the two complex conjugate branch points where E_1 and E_2 coalesce.[6] We discern two Riemann sheets with values E_1 and E_2 and square root branch points at the zeros of the discriminant. Encircling either zero once brings us from one sheet to the other.

Somewhat less trivial is the case $N = 3$. For illustration we proceed now by first determining the resultant. The derivative of Eq.(7) with respect to E yields a quadratic equation. Its solution is inserted into Eq.(7). After rearranging terms we thus obtain for the resultant

$$R(\lambda) = c_1^2 c_2^2 + 18 c_1 c_2 c_3 - 4 c_2^3 - 4 c_1^3 c_3 - 27 c_3^2$$

(9)

the zeros of which are the exceptional points. As $R(\lambda)$ is of sixth order there are three pairs of complex conjugate zeros. In turn, using the explicit expressions[8] for

W.D. HEISS[†] AND W.-H. STEEB[‡]

the solution of Eq.(7) we obtain the three roots

$$E_1(\lambda) = x_1(\lambda) + x_2(\lambda) - \frac{c_1(\lambda)}{3}$$

$$E_2(\lambda) = \alpha^2 x_1(\lambda) + \alpha x_2(\lambda) - \frac{c_1(\lambda)}{3} \qquad (10)$$

$$E_3(\lambda) = \alpha x_1(\lambda) + \alpha^2 x_2(\lambda) - \frac{c_1(\lambda)}{3}$$

where

$$x_{1,2}(\lambda) = \sqrt[3]{r(\lambda) \pm \sqrt{r^2(\lambda) + q^3(\lambda)}}$$

$$q(\lambda) = \frac{c_2(\lambda)}{3} - \frac{c_1^2(\lambda)}{9}$$

$$r(\lambda) = \frac{c_2 c_1 - 3c_3}{6} - \frac{c_1^3}{27}$$

$$\alpha = \exp\frac{2i\pi}{3}.$$

Here we recognise the discriminant $D(\lambda) = r^2(\lambda) + q^3(\lambda)$ and we can verify the identity $R(\lambda) = -108D(\lambda)$. Note that $D(\lambda)$ is non-positive for real λ as the $E_n(\lambda)$ are real. Since even in three dimension the general case lacks transparence we present an illustrative special example.

We consider the problem

$$H(\lambda) = \begin{pmatrix} 0 & 0 & 0 \\ 0 & 1 & 0 \\ 0 & 0 & 2 \end{pmatrix} + \lambda \begin{pmatrix} 0 & 1 & 0 \\ 1 & 0 & c \\ 0 & c & 0 \end{pmatrix} \qquad (11)$$

with c real. From Eq.(1) we obtain

$$-E^3 + 3E^2 + E(c^2\lambda^2 + \lambda^2 - 2) - 2\lambda^2 = 0 \qquad (12)$$

and from Eq.(2)

$$E^2 - 2E - \frac{1}{3}(c^2\lambda^2 + \lambda^2 - 2) = 0. \qquad (13)$$

Solving Eq.(13) and inserting the solution into Eq.(12) yields a third order equation in λ^2 which can therefore be solved explicitly. Using the notation

$$A = -c^3 + c^2 - c + 1$$

$$B = 10c^4 - 52c^2 + 10$$

$$C = 3c^3 - 15c^2 - 15c + 3$$

we find for the exceptional points

$$\lambda_1(c) = \frac{\sqrt{A^{1/3}C(-i\sqrt{3}-1) + B - 3A^{2/3}(c^2 + 6c + 1)(i\sqrt{3}-1)}}{2\sqrt{2c^2 + 2(c^2 + 1)}} \tag{14a}$$

and

$$\lambda_2(c) = -\lambda_1(c), \quad \lambda_3(c) = \bar{\lambda}_1(c), \quad \lambda_4(c) = -\bar{\lambda}_1(c) \tag{14b}$$

and

$$\lambda_5(c) = \frac{\sqrt{A^{1/3}C + B/2 - 3A^{2/3}(c^2 + 6c + 1)}}{2\sqrt{c^2 + 1}(c^2 + 1)}, \quad \lambda_6(c) = \bar{\lambda}_5(c). \tag{14c}$$

In this case the exceptional points occur symmetrical with respect to the imaginary axis. This is a consequence of the particular form of H_1 and is true in N dimensions: if H_1 is tridiagonal with vanishing diagonal, the determinants in Eqs.(1) and (2) are functions of λ^2. As a further consquence, for $N(N-1)/2$ odd, one pair of exceptional points must lie on the imaginary axis as is the case for λ_5 and λ_6 above

There are limiting cases for which the three sheets collaps into two plus one which is unconnected by the confluence of some of the exceptional points. For $c \to \infty$ we obtain

$$\lambda_1(c) \to \frac{\sqrt{2}}{c}, \quad \lambda_5 \to \frac{i}{2c}$$

while for $c = 1$ the three exceptional points in the upper plane merge into one, *viz.*

$$\lambda_1 = \lambda_4 = \lambda_5 = \frac{i}{\sqrt{2}}$$

with the corresponding pattern in the lower plane

$$\lambda_2 = \lambda_3 = \lambda_6 = -\frac{i}{\sqrt{2}}.$$

The latter case corresponds to the energies

$$E_1(\lambda) = -\sqrt{2\lambda^2 + 1} + 1, \quad E_2(\lambda) = 1, \quad E_3(\lambda) = \sqrt{2\lambda^2 + 1} + 1$$

which clearly demonstrates that only two levels are connected and one (E_2) is disconnected.

4. Physical significance of exceptional points

In the following we demonstrate that coalescence of two levels for complex values of λ implies avoided level crossing for real values of λ for the two levels concerned. We further argue below that the local behaviour of an N-dimensional problem is like that of a two dimensional problem if the exceptional point is isolated from further singularities. This can be generalised in that a few closely lying singularities behave locally like a three or four dimensional matrix problem.

We assume that the two levels $E_n(\lambda)$ and $E_{n+1}(\lambda)$ coalesce at the points λ_c and at its complex conjugate $\bar{\lambda}_c$. We define the functions

$$g(\lambda) = \frac{1}{2}(E_n(\lambda) + E_{n+1}(\lambda))$$

$$f(\lambda) = \frac{E_n(\lambda) - E_{n+1}(\lambda)}{2\sqrt{(\lambda - \lambda_c)(\lambda - \bar{\lambda}_c)}}. \tag{15}$$

Both functions are regular at λ_c and $\bar{\lambda}_c$, while they have singularities (square root branch points) at the exceptional points where E_n and E_{n+1} are connected to levels with labels different from $n+1$ and n, respectively. Rewriting Eq.(15) we obtain

$$E_n(\lambda) = g(\lambda) + \sqrt{(\lambda - \lambda_c)(\lambda - \bar{\lambda}_c)}f(\lambda)$$

$$E_{n+1}(\lambda) = g(\lambda) - \sqrt{(\lambda - \lambda_c)(\lambda - \bar{\lambda}_c)}f(\lambda). \tag{16}$$

While this is an identity, the expression reveals explicitly the connectedness of the two levels and, since the functions f and g are regular at λ_c and $\bar{\lambda}_c$, the singularity structure is explicitly displayed.

The effect of that singularity upon the spectrum for real λ means avoided level crossing if it is isolated. The minimum spacing between the two levels occurs at the (real) value of λ which obeys the condition

$$f(\lambda)(\lambda - \Re\lambda_c) + |\lambda - \lambda_c|^2 f'(\lambda) = 0 \tag{17}$$

where $\Re\lambda$ denotes the real value of λ and f' the derivative of f. The second term of Eq.(17) is small if the derivative of f is small or if λ_c is near to the real axis; the derivative of f is expected to be small, if the nearest singularity of f is not too close. We thus find under these conditions

$$\lambda_{\min} \approx \Re\lambda_c \tag{18}$$

which means that avoided level crossing occurs just at $\Re\lambda$. For the value of the spacing we find

$$\Delta \approx 2f(\Re\lambda) \cdot \Im\lambda_c \tag{19}$$

with $\Im\lambda$ denoting the imaginary part of λ. While this relation indicates an approximate linear dependence of the spacing on $\Im\lambda_c$ it is not as useful as Eq.(18) as the proportionality constant $f(\Re\lambda)$ is not known. Note that these relations become exact for $N = 2$ as f is constant in this case. In this way we have shown that the structure of a two by two problem prevails for two adjacent levels of an $N \times N$ problem.

For $N > 2$, $f'(\Re\lambda_c)$ can be large due to a near singularity. This could affect Eq.(18). The function $f(\lambda)$ still carries all the singularities of E_n and E_{n+1} where energy levels with labels different from n and $n+1$ are connected. If they are close to λ_c, Eq.(18) could be invalidated. In this context one must be aware of the Riemann sheet structure: singularities are close to each other only if they lie in the same sheet. Because the separation by a number of sheets does not allow a short path between the two singularities, it is sufficient to take into account only a few more adjacent levels to incorporate the effect of nearby singularities. How to do this for three or four levels is discussed in detail elsewhere.[9]

5. Nonlinear differential equations

A set of nonlinear differential equations has been derived[10] for the levels and matrix elements taken with the eigenfunctions $\psi_n(\lambda)$. They read

$$\frac{dE_n}{d\lambda} = p_n$$

$$\frac{dp_n}{d\lambda} = 2 \sum_{m \neq n} \frac{V_{mn} V_{nm}}{E_n - E_m}$$

$$\frac{dV_{mn}}{d\lambda} = \sum_{k \neq m,n} V_{mk} V_{kn} \left(\frac{1}{E_m - E_k} + \frac{1}{E_n - E_k} \right) - V_{mn} \frac{p_m - p_n}{E_m - E_n}$$

where we used the notation

$$p_n(\lambda) = \langle \psi_n(\lambda)|H_1|\psi_n(\lambda)\rangle, \quad V_{mn}(\lambda) = \langle \psi_m(\lambda)|H_1|\psi_n(\lambda)\rangle, \ m \neq n.$$

It is obvious that the exceptional points must be singularities of this system. This was in fact demonstrated[11] in a singular point analysis of the system. A so-called Painlevé test[12] reveals that the behaviour of the matrix elements is dominated by $\sim (\lambda - \lambda_c)^{-1/2}$ at the points λ_c, where the energy levels behave as in Eq.(16). This particular singular behaviour which is due to the eigenstates was found[13] also for solutions of the Schrödinger equation for a single particle potential which corresponds, in our context, to an infinite dimensional problem.

References

1. J. von Neumann and E.P. Wigner, Z. Phys. **30** (1929), 467
2. G. Casati (editor) *Chaotic behaviour in quantum systems*, New York: Plenum 1985
3. T. Kato *Perturbation Theory of Linear Operators*, Springer 1966
4. C.M. Bender and T.T. Wu, Phys. Rev. **D7** (1973), 1620
5. P.E. Shanley, Ann. Phys. (N.Y.) **186** (1988), 292
6. W.D. Heiss and A.L. Sannino, J. Phys. A: Mathematical and General, **23** (1990), 1167; W.D.Heiss and A.L. Sannino, Phys. Rev. **A43** (1991), 4159; W.D. Heiss and A.A. Kotzé, Phys. Rev **A44** (1991), 2403
7. A. Mostowsky and M. Stark *Introduction to Higher Algebra*, Pergamon Press 1964
8. M. Abramowitz and I.A. Stegun *Handbook of Mathematical Functions*, Dover Publ. Inc., New York 1965
9. W.D. Heiss and W.-H. Steeb, J. Math. Phys. **32** (1991), 3003
10. P. Pechukas, Phys. Rev. Lett. **51** (1983), 943; T. Yukawa, Phys. Rev. Lett. **54** (1985), 1883; W.-H. Steeb *Hilbert Spaces, Generalised Functions and Quantum Mechanics*, BI-Wissenschaftsverlag, Mannheim, 1991.
11. W.-H. Steeb and W.D. Heiss, Phys. Lett. **A152** (1991), 339; W.-H. Steeb and W.D. Heiss, Physica Scripta. **47** (1993), 321

12. W.-H Steeb and N. Euler, *Nonlinear evolution equations and Painlevé test*, World Scientific,
     Signapore, 1988; W.-H. Steeb, *Problems in theoretical physics* II, BI-Wissenschaftsverlag,
     Mannheim, 1990
13. W.D. Heiss, Nucl.Phys. **A144** (1970), 417

TYPE-II INTERMITTENCY IN THE PRESENCE OF ADDITIVE AND MULTIPLICATIVE NOISE

H. HERZEL
Institute of Theoretical Physics
Technical University Berlin
Hardenbergstr. 36
D-10623 Berlin
Germany

and

F. ARGOUL AND A. ARNEODO
Centre de Recherche Paul Pascal
Domaine Universitaire
33405 Talence Cedex
France

Abstract. Effects of noise on the transition from periodic to intermittent chaotic behaviour via a subcritical Hopf bifurcation are studied. For this purpose numerical simulations of a three-dimensional map and of a Hopf normal form map are carried out.

The resulting histograms and mean lengths of laminar episodes are compared with analytical results from a one-dimensional stochastic model with non-vanishing probability flow.

A power-law decrease of the length of the laminar phases is found for additive noise whereas multiplicative fluctuations lead to a drastic increase. Using asymptotic formulae this effect is explained as a suppression of hyperbolic growth due to noise.

1. Introduction

Pomeau and Manneville introduced the notion of intermittency as a specific route to weak turbulence [1] and stimulated many theoretical, numerical and experimental studies during the past few years (see e.g. [2] - [11]). In termittency is characterized by a local instability of a limit cycle which can be either a saddle-node bifurcation (type-I intermittency), or a subcritical Hopf bifurcation (type-II intermittency) or, a subcritical period-doubling bifurcation (type-III intermittency). In addition, a global nonlinear mechanism, e.g. strange attractor like behaviour, is necessary to ensure the reinjection of the trajectories into the neighbourhood of the weakly unstable limit cycle. Then, when varying a control parameter, a continuous transition is observed from a periodic regime to nearly periodic oscillations (termed laminar phases) which are interrupted by short turbulent bursts.

The statistics of the laminar episodes plays a central role for the quantitative description of intermittency since it is easily accessible in experiments. When observations are compared with theory it is essential to understand the role of fluctuations. In particular, it is now widely accepted that beside additive noise also parametric (or multiplicative) perturbations play a crucial role in many fields, e.g. electronics, optics, chemical kinetics and population dynam ics [12] - [16].

99

Z. Haba et al. (eds.), Stochasticity and Quantum Chaos, 99–113.

The aim of this paper is to study the effects of noise on type-II intermittency which is undoubtedly the least popular type since only recently numerical [17, 18] and real experiments [7, 10, 11] identifying this route were published.

In section 2 the main results of numerical experiments on a three-dimensional map modelling type-II intermittency are presented. As intuitively expected, additive noise decreases the mean laminar times, while in the case of parametric fluctuations an increase is observed. Such "stabilizing" effects of fluctuations are always of particular interest [19, 20].

The numerical observations can be explained with the aid of analytical calculations given in the sections 3 to 5. A normal form of a subcritical Hopf bifurcation with appropriate boundary conditions is analyzed, and the mean laminar time is derived directly from the stationary probability distribution.

Section 6 is devoted to a comparison of the analytical results and simulations of a stochastic discrete map related to the Hopf normal form. In a concluding section the applicability of some of our results to type-III in termittency is emphasized. The equivalence of different approaches to obtain the mean laminar time is shown in an Appendix.

2. Numerical Simulations of a Three-Dimensional Map

Often, only formal models of type-II intermittency have been studied which are based on a subcritical Hopf normal form together with an artificially added rein jection mechanism [1, 21]. In Ref. [18] a three-dimensional dissipative map is proposed to model type-II intermittency in close analogy to a Poincaré section of a periodically driven third-order nonlinear oscillator [17]:

$$
\begin{aligned}
x_{n+1} &= x_n + y_n \\
y_{n+1} &= y_n + z_n \\
z_{n+1} &= (1 - \eta)z_n - \nu y_n - \mu x_n - k_1 x_n^2 - k_2 y_n^2 - k_3 x_n y_n - k_4 x_n z_n - k_5 x_n^2 z_n
\end{aligned}
\tag{1}
$$

Fig. 1 displays the alternation of laminar episodes and bursts generated by model (1).

It was found that the homoclinic reinjection process is essentially one-dimensional (see Fig. 2) which has consequences for the mean length of the laminar phases [17].

Additive fluctuations are modelled by the insertion of a term $h\xi_n$ into the third equation of Eqs. 1. Here, the ξ_n are independent Gaussian pseudo-random numbers with zero mean and a variance equal to unity. It can be seen from Fig. 3a that the mean length of laminar phases $< n >$ decreases with increasing noise intensity, and it turns out to be independent of ϵ for sufficiently large noise amplitudes h. In that region a power-law

$$
< n > \sim h^{-1}
\tag{2}
$$

is approximately fulfilled.

The effect of a fluctuating control parameter ϵ is quite different (see Fig. 3b). Beyond some cut-off value a drastic increase of the time $< n >$ with noise strength h is observed.

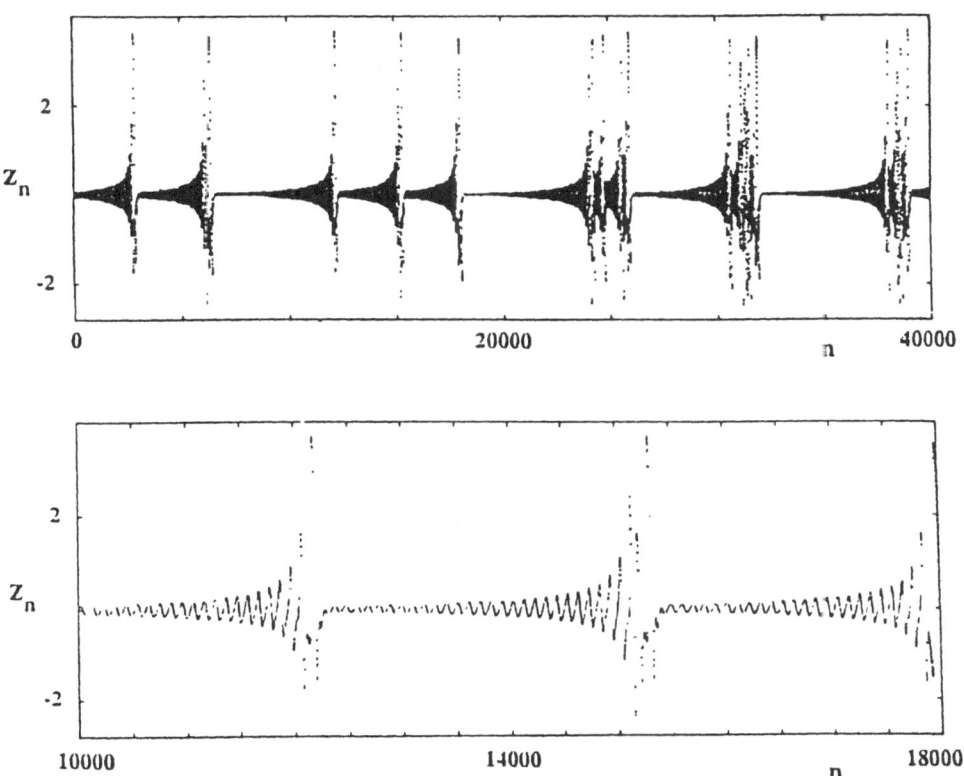

Fig. 1.   Characteristic intermittent realization from Eqs. 1 (parameter values are given in Ref. [17]). The distance from the Hopf bifurcation at μ_c is $\epsilon = \mu - \mu_c$. In this figure ln $\epsilon = -11$ is chosen. In the lower plot an enlargement of the realization is given.

This remarkable response to fluctuations motivated us to study a stochastic version of the type-II intermittency normal form in detail. The applicability of a subcritical Hopf normal form is emphasized in Fig. 2a. In the stochastic model the global reinjection mechanism of Eqs. 1 is replaced by appropriate boundary conditions and a reentering flux.

3. Derivation of a Stochastic Model

Earlier papers on intermittency in the presence of noise are concerned either with type-I [2, 3, 4, 22, 23, 24] or type-III intermittency [25, 26, 27] and are adressed to the influence of additive noise. As main tools stationary probability distributions and mean first passage times are derived in these papers. Below, we concentrate on the

POINCARE MAP   REINJECTION DISTRIBUTION

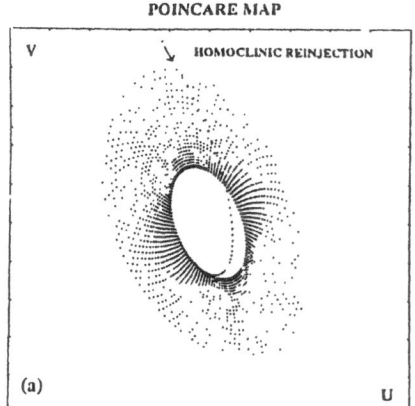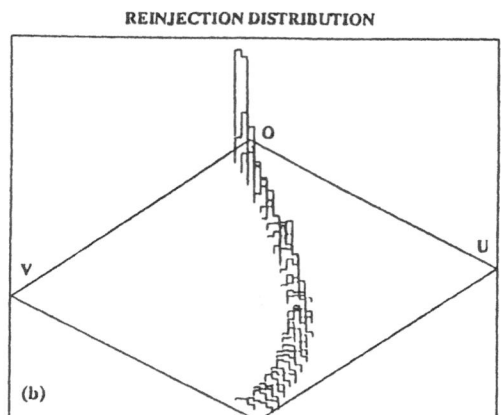

Fig. 2.   (a) Projection of the Poincaré map onto the local two-dimensional unstable manifold of the origin.   Only a finite number of iterates of the mapping (1) are considered in order to visualize the homoclinic reinjection and a spiraling-out laminar episode.   (b) Histogram of the reinjection distribution inside the local two-dimensional unstable manifold of the origin 0.   Same model parameters as in Fig. 1.

calculation of stationary probability densities from Fokker-Planck equations since the mean length of the laminar phases can be obtained easily from the stationary probability flow. The equivalence of this approach to the mean first passage time calculus is exemplified in the Appendix. The case of multiplicative noise which exhibits nontrivial effects (see Fig. 3b) will be discussed first and more extensively.

As already mentioned in the introduction, type-II intermittency is connected with a subcritical Hopf bifurcation of a limit cycle. Therefore, oscillations with slowly varying amplitudes can be observed near the bifurcation (see Figs. 1a and 2a). Using polar coordinates the following normal form of the amplitude equ ation is obtained after phase averaging [1, 21].

$$r_{n+1} = (1 + \epsilon)r_n + ar_n^3 \qquad (a \geq 0) \qquad (3)$$

For sufficiently small ϵ this discrete version can be approximated by a continuous equation as in Refs. [3, 21, 24]. If the control parameter ϵ is perturbed by Gaussian white noise the fol lowing stochastic differential equation results:

$$\frac{dr}{dt} = \epsilon r + ar^3 + hr\xi(t) \qquad (4)$$

$$< \xi(t) >= 0 \quad ; \quad < \xi(t)\xi(t + \tau) >= \delta(\tau)$$

Clearly, Eq. 4 can serve only as a local model since there is no limitation of growth. From global models like Eqs. 1 it is known that for overcritical amplitudes short

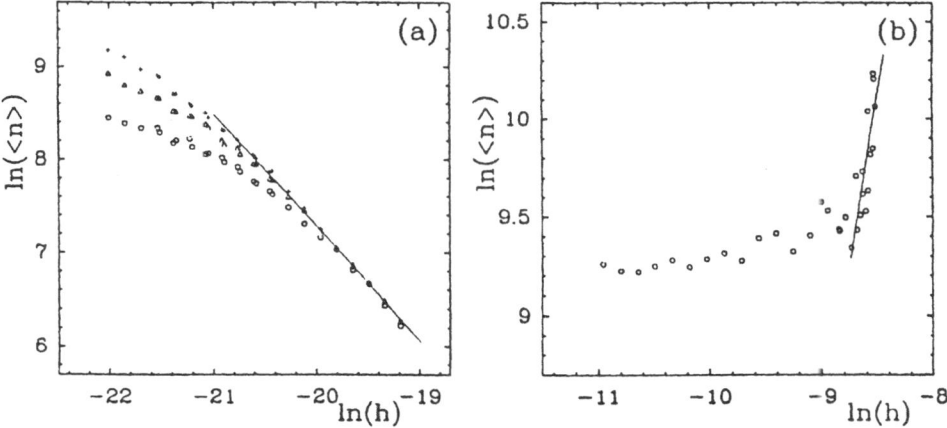

Fig. 3. The average length of the laminar episodes $< n >$ versus the amplitude of the noise h in log-log scales (parameters are given in Ref. [17]). (a) additive noise. circles: $\ln \epsilon = -12$; triangles: $\ln \epsilon = -13$; crosses: $\ln \epsilon = -14$ (b) multiplicative noise. $\ln \epsilon = -12$; The solid line is drawn to guide the eye.

chaotic bursts occur which lead to a reinjection near the origin. Such a dynamics is "translated" to our one-dimensional model (4) as follows: At $r = R$ an absorbing boundary is taken and a reentering flux $F(r)$ is included. The other boundary at $r = 0$ is an inaccessible "built-in" boundary [28].

With respect to the flux $F(r)$ we distinguish two cases:

(a) pointwise reinjection at $r = A$

(b) equidistributed reinjection into the interval $(0, B)$.

The first case offers the opportunity to average afterwards over arbitrary reinjection distributions. The second case is motivated by the simulation results of model (1) where an essentially one-dimensional reinjection was observed.

Thus, a well-defined stochastic process is obtained which contains despite several simplifications the main features of systems near the type-II intermittency threshold.

4. Stationary Probability Density

Using the Stratonovich prescription [12] the Langevin equation (4) corresponds to the following Fokker-Planck equation taking into account the reentering flux $F(r)$ which is specified below.

$$\frac{\partial}{\partial t} P(r,t) = -\frac{\partial}{\partial r}[(\epsilon r + ar^3 + \frac{h^2}{2}r)P] + \frac{h^2}{2}\frac{\partial^2}{\partial r^2}[r^2 P] + F(r) \qquad (5)$$

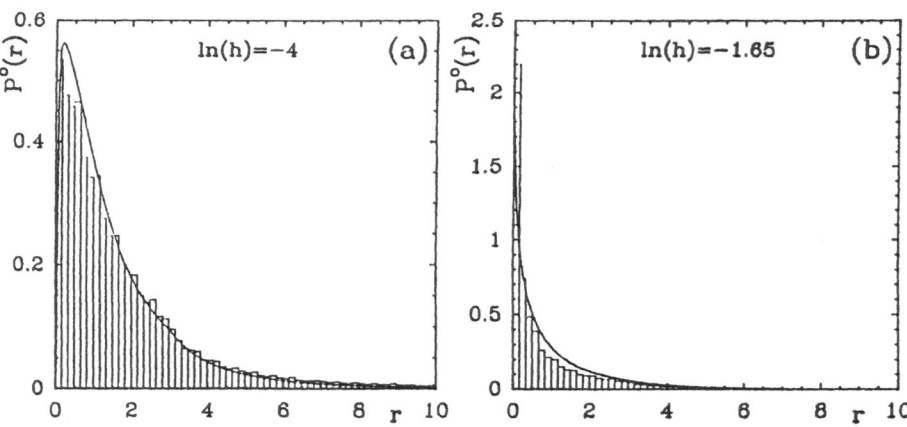

Fig. 4.   Histograms of sojourn times from the two-dimensional normal form map (26) (ln $\epsilon = -4$).
For comparison the densities from the numerical integration of Eq. 10 are presented (full lines).
Note the different vertical scales.

In the following we are interested in the stationary solution $P^0(r)$. Integrating once
we obtain

$$(\epsilon r + ar^3 + \frac{h^2}{2}r)P^0 - \frac{h^2}{2}\frac{\partial}{\partial r}[r^2 P^0] = \int_0^r F(u)\,du \equiv I(r). \tag{6}$$

Now the flux $F(r)$ will be specified along the lines of the preceding section.   (a)
pointwise reinjection:

$$F(r) = G\,\delta(r - A) \tag{7}$$
$$I(r) = G\,\Theta(r - A)$$

(b) uniform reinjection into $(0, B)$:

$$F(r) = F\,\Theta(B - r) \tag{8}$$
$$I(r) = F\,r \quad \text{for} \quad 0 < r < B$$
$$\qquad = F\,B \equiv G \quad \text{for} \quad B \leq r \leq R$$

We have to note that the integration constant arising from the integration of Eq. 5
vanishes since no probability flow through the boundary r=0 is possible. At r=R we

have an absorbing boundary with

$$P^0(R) = 0. \tag{9}$$

Integration of Eq. 6 gives then the following stationary probability density:

$$P^0(r) = \frac{2}{h^2} r^{\frac{2\epsilon}{h^2}-1} \exp(\frac{a}{h^2}r^2) \int_r^R I(u) u^{-\frac{2\epsilon}{h^2}-1} \exp(-\frac{a}{h^2}u^2) \, du \tag{10}$$

The free constant G in $I(u)$ has to be determined from the normalization condition

$$\int_0^R P^0(r) \, dr = 1. \tag{11}$$

For $\epsilon \leq 0$ the distribution is not normalizable, but we are mainly interested in positive values of ϵ corresponding to intermittent behaviour.

In section 6 the above distribution is compared with histograms from simulations of a discrete stochastic map. The shape of the density (10) changes drastically at $h = \sqrt{2\epsilon}$ which was termed "noise-induced transition" [16]. We postpone the discussion of the effect of this transition on the mean length of laminar episodes to the next section.

Now we turn to the case of additive noise which is surprisingly somewhat more complicated since additive noise sources exhibit not a priori radial symmetry as fluctuations of the parameter ϵ do. However, if we take a Hopf normal form in Cartesian coordinates and add independent noise sources of the same intensity a one-dimensional Fokker-Planck equation can be derived [29].

$$\frac{dx}{dt} = \omega y + \epsilon x + ax(x^2 + y^2) + h\xi_1(t) \tag{12}$$

$$\frac{dy}{dt} = -\omega x + \epsilon y + ay(x^2 + y^2) + h\xi_2(t)$$

$$< \xi_i(t) >= 0 \quad ; \quad < \xi_i(t)\xi_j(t+\tau) >= \delta_{ij}\delta(\tau)$$

Introducing polar coordinates and making use of the radial symmetry of the stationary probability density the following equation results [29]:

$$-\frac{\partial}{\partial r}[(\epsilon r + ar^3 + \frac{h^2}{2r})P^0] + \frac{h^2}{2}\frac{\partial^2}{\partial r^2}P^0 + F(r) = 0 \tag{13}$$

The term $\frac{h^2}{2r}$ arises from the transformation to polar coordinates. In the same manner as for multiplicative noise the stationary probability density can be derived:

$$P^0(r) = \frac{2}{h^2} r \exp(\frac{\epsilon r^2 + \frac{a}{2}r^4}{h^2}) \int_r^R \frac{I(u)}{u} \exp(-\frac{\epsilon u^2 + \frac{a}{2}u^4}{h^2}) \, , du \tag{14}$$

Again, the normalization condition gives the value of the probability flow G.

5. Mean Length of Laminar Phases

First we discuss the deterministic time T_A to reach the boundary R from an injection point A in the absence of fluctuations. For $a = 0$ we find exponential growth with

$$T_A = \frac{1}{\epsilon} \ln \frac{R}{A} \qquad \text{for} \qquad a = h = 0, \tag{15}$$

whereas in the case $\epsilon = 0$ hyperbolic growth occurs with

$$T_A = \frac{1}{2a} \left(\frac{1}{A^2} - \frac{1}{R^2} \right) \qquad \text{for} \qquad \epsilon = h = 0. \tag{16}$$

In general, there is a cross-over from exponential to hyperbolic growth around $r = \sqrt{\frac{\epsilon}{a}}$, and integration of Eq. 4 gives

$$T_A = \frac{1}{2\epsilon} \ln \frac{1 + \dfrac{\epsilon}{aA^2}}{1 + \dfrac{\epsilon}{aR^2}} \qquad \text{for} \qquad h = 0. \tag{17}$$

The standard approach for the calculation of the mean length of laminar episodes in the presence of noise is the computation of the mean first passage time (see Appendix). For our problem it is more convenient to use the results of the preceding section. There, we found a constant probability flow for $r > A$ (or $r > B$, respectively). Thus, the probability that a realization leaves the intervalat R per unit time is just G. The mean sojourn time is, therefore, the inverse of G [24]. Keeping in mind that G is determined by the normalization condition (11) the following formulaes for the mean laminar times result for multiplicative noise:

$$\text{(a)} \quad T_A = \frac{2}{h^2} \int_0^R v^{\frac{2\epsilon}{h^2}-1} \exp(\frac{a}{h^2}v^2) \int_v^R \Theta(u - A)\, u^{-\frac{2\epsilon}{h^2}-1} \exp(-\frac{a}{h^2}u^2)\, du\, dv \tag{18}$$

$$\text{(b)} \quad \bar{T}_B = \frac{2}{h^2} \int_0^R v^{\frac{2\epsilon}{h^2}-1} \exp(\frac{a}{h^2}v^2) \int_v^R H(u)\, du\, dv \tag{19}$$

$$\text{with} \qquad H(u) = \frac{1}{B} u^{-\frac{2\epsilon}{h^2}} \exp(-\frac{a}{h^2}u^2) \quad \text{for} \quad 0 \le u \le B$$

$$= u^{-\frac{2\epsilon}{h^2}-1} \exp(-\frac{a}{h^2}u^2) \quad \text{for} \quad B < u \le R$$

Here, the index A marks the pointwise reinjection and the index B together with a bar on top of T marks the uniform reinjection into $(0, B)$. These results may be checked for consistency since averaging T_A over the interval $(0, B)$ should give $\bar{T}_B$. This identity can indeed be proven (see Appendix). For additive noise analogous formulaes can be derived easily from Eq. 14 which are omitted here.

Now the question arises how the mean laminar times depend on the noise intensity h. Of course, the integrals (18) and (19) can be easily calculated numerically

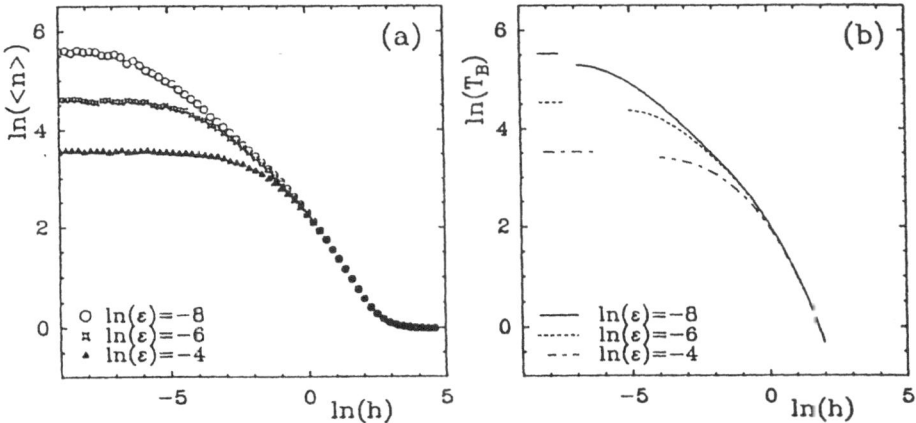

Fig. 5.   Mean length of laminar phases versus noise amplitude for additive noise. (a) Simulations of the map (26) (b) Numerical integration of the analytic expression in Eq. 19. The horizontal lines correspond to the deterministic limit.

(which was done in the Figs. 5b and 6b), but we intend also the derivation of simple analytical expressions using the following approximations: For small u (or v), i.e. for $u(\text{or } v) \leq \frac{h}{\sqrt{a}} \equiv z$, the exponential function in Eq. 18 is replaced by unity, while for large u the first term of an asymptotic expansion is used:

$$\int_{v}^{R} u^{-\frac{2\epsilon}{h^2}-1} \exp(-\frac{a}{h^2}u^2)\,du \approx \frac{h^2}{2a}v^{-\frac{2\epsilon}{h^2}-2} \exp(-\frac{a}{h^2}v^2) \tag{20}$$

$$\text{if} \quad v > \frac{h}{\sqrt{a}} \equiv z \quad \text{and} \quad R \quad \text{large}$$

These approximations are sufficient to derive a rather simple expression for T_A in the case of multiplicative noise:

$$T_A = \frac{1}{\epsilon}\ln\frac{z}{A} + \frac{1}{2a}\left(\frac{1}{z^2} - \frac{1}{R^2}\right) + \frac{1}{2\epsilon e} \tag{21}$$

$$\text{for} \quad A < z \leq R$$

For $A > z = \frac{h}{\sqrt{a}}$, a similar result can be obtained which is omitted here. Furthermore, averaging T_A over $(0, B)$ gives the corresponding approximations of $\bar{T}_B$. Formula (21) resembles the Eqs. 15 and 16 which describe the deterministic times of exponential or hyperbolic growth processes. Eq. 21 suggests, therefore, the following interpretation: For $r < z = \frac{h}{\sqrt{a}}$ exponential growth dominates, whereas for $r > z$ hyperbolic behaviour is found. On the other hand, the deterministic cross-over is found at $r = \sqrt{\frac{\epsilon}{a}}$. Thus, we can conclude immediately that for $h > \sqrt{\epsilon}$ the "exponential region" is enlarged and, consequently, the time T_A increases with noise intensity h. The limiting case of large h can be obtained directly from Eq. 18 if we assume

$$\exp(\frac{a}{h^2}v^2) = \exp(-\frac{a}{h^2}u^2) = 1 \qquad (22)$$

Then, the hyperbolic growth is suppressed totally and we find

$$T_A = \frac{1}{\epsilon} \ln \frac{R}{A} \qquad (23)$$

In this way our asymptotic expansions suggest an increase of T_A from the deterministic value in Eq. 17 to the limit (23). Such an increase is found indeed by direct integration of Eq. 19 (see Fig. 6b in the next section).

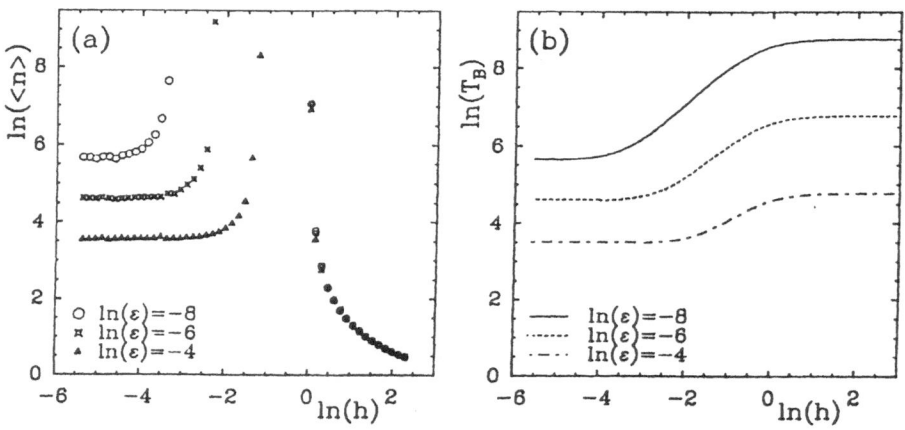

Fig. 6.   The same as in Fig. 5 for multiplicative noise.

It is worth noting that Eq. 23 holds exactly in the linear case (i.e. $a = 0$) for arbitrary noise intensities. Thus, multiplicative noise neither accelerates nor delays the exponential growth on an average. Therefore, the "noise-induced transition" at $\epsilon = \frac{h^2}{2}$, which can be found also for $a = 0$, is not sufficient for an increase of T_A. Only a relevant nonlinearity (i.e. $\sqrt{\frac{\epsilon}{a}} < R$) leads to the described increase of T_A.

Also for additive noise the corresponding approximations lead to really elucidating results. The numerical simulations in section 2 suggest a power-law $T_A \sim h^{-1}$ if the time is independent of ϵ.

Using asymptotic expansions as above we find for $\epsilon = 0$:

$$T_A = \frac{\tilde{z}^2 - A^2}{h^2} + \frac{1}{2a}\left(\frac{1}{\tilde{z}^2} - \frac{1}{R^2}\right) + \frac{\exp(-1)}{h\sqrt{8a}} \qquad (24)$$

$$\text{for} \quad A < \left(\frac{2h^2}{a}\right)^{\frac{1}{4}} \equiv \tilde{z} \leq R$$

The first term corresponds to random walk for $r < \tilde{z}$ whereas the second term describes a hyperbolic growth process. Having $\tilde{z}^4 = \frac{2h^2}{a}$ in mind it can be seen clearly that the dominant terms in Eq. 24 confirm the simulation result $T_A \sim h^{-1}$. In the limiting case of large h a pure random walk results

$$T_A = \frac{R^2 - A^2}{2h^2}, \qquad (25)$$

in contrast to the case of multiplicative noise where the exponential growth persists.

6. Comparison of Analytical Results with Numerical Simulations

In order to test the predictions of the theoretical model above we study in this section the complex normal form map

$$z_{n+1} = (1 + \epsilon)\, e^{i\omega}\, z_n + a\, e^{i\omega}\, z_n |z_n|^2. \qquad (26)$$

Additive noise is simulated by the insertion of independent Gaussian pseudo-random terms $h\xi_n$ to the real and imaginary part of z_n with

$$< \xi_n >= 0 \quad ; \quad < \xi_n \xi_m >= \delta_{nm}$$

Similarly, multiplicative noise is modelled by the substitution of ϵ by $\epsilon + h\xi_n$. The map (26) is iterated until $|z_n|$ exceeds R which is taken as the end of a laminar phase, and then a new initial condition with $0 < z \leq B$ is randomly chosen. With the simulation results of map (1) in mind (one-dimensional reinjection) the reentry points are not uniformly distributed in a disk as proposed in Ref. [1] but they lie rather on a line. Thus, our numerical procedure corresponds to the reinjection mechanism (b) of our analytical calculations. First we compare histograms from model (26) with the stationary probability density from Eq. 10 (full lines in Fig. 4). There is a fairly good agreement between the simulations of the mapping and the analytic results from the continuous model (4). Moreover, the stabilizing effect of multiplicative noise is visible, since for higher noise intensity the density is more concentrated around the origin.

Fig. 5 shows the dependance of the mean length of laminar episodes on the noise strength in the case of additive noise. For $h < 1$, there is an excellent agreement between the simulation of the map (26) in Fig. 5a and the analytical results in Fig. 5b. For large noise intensities ($h > 1$) it is no longer justified to replace the stochastic map by a Langevin equation and, consequently, the curves differ.

The coincidence of simulations and theory is less satisfactory in the case of multiplicative noise. The increase in Fig. 6a is steeper than in Fig. 6b. The disagreement for $h > 1$ is not unexpected since the continuous approach is not applicable for such large noise intensities.

It is worth mentioning that the deterministic limit and the asymptotic formula (23) describe the main features of the curves in Fig. 6b. There is indeed a step wise increase of T_A for $h \sim \sqrt{\epsilon}$. The height of the step diverges for $\epsilon \rightarrow 0$ and, therefore, we expect a better agreement between simulations and theory for smaller values of ϵ and h. However, in any case, the increase of the mean laminar time with multiplicative noise was reproduced by the analytic approach.

7. Concluding Remarks

Most calculations of stationary probability distributions are concerned with "natural boundary conditions" and, therefore, with a vanishing probability flow [16, 30, 31]. The transition to chaos via intermittency leads to peculiar stochastic models with non-vanishing flow and a reentering flux. We note that similar approaches play a role in the stochastic theory of nucleation [32] and for nonstationary processes of lasers [33].

After reducing the intermittency type-II transition to a one-dimensional problem the stationary probability density was derived. In a convenient way the mean length of laminar phases was obtained directly from the probability flow. Using asymptotic expansions several effects could be understood. For additive noise a power-law decrease of the length of laminar episodes results which was found numerically and from analytical calculations.

The most interesting effect was a sudden increase of the laminar times due to a fluctuating order parameter. It can be interpreted as a suppression of the hyperbolic acceleration due to noise. Consequently, a suppression of a hyperbolic damping (i.e. $a < 0$) should lead to a decrease of the mean first passage time. This effect was found indeed [34]. Finally, it is worth mentioning that a good agreement of the analytical theory with several numerical simulations was obtained.

Although all simulations above are related to type-II intermittency we emphasize that the analytical results on multiplicative noise are also valid for type-III intermittency since the corresponding normal form [21]

$$x_{n+2} = (1 + 2\epsilon)x_n + \beta x_n^3 \tag{27}$$

resembles clearly Eq. 3. In both cases, a one-dimensional reinjection is appropriate. In particular, it is absolutely neccessary to study multiplicative perturbations of the map (27) since the transition from the original map

$$x_{n+1} = -(1 + \epsilon)x_n + \alpha x_n^2 + \beta' x_n^3 + h\xi_n \tag{28}$$

to the normal form (27) transforms originally additive noise into multiplicative fluctuations:

$$x_{n+2} = (1 + 2\epsilon - 2\alpha h\xi_n)x_n + \beta x_n^3 + h(\xi_n + \xi_{n+1}) + \text{h.o.t.} \tag{29}$$

$$\text{with} \quad \beta = -2(\alpha + \beta')$$

Thus, the described effects could be observed in careful experiments of type-III intermittency as well.

Appendix

There are (at least) three approaches to compute the mean length of laminar episodes for a uniform reinjection into $(0, B)$:
i) averaging T_A from Eq. 18 over the interval $(0, B)$
ii) calculating $\bar{T}_B$ from Eq. 19
iii) averaging the mean first passage time $M(A)$ over $(0, B)$.
The aim of the Appendix is to point out the equivalence of these three possibilities. The mean first passage time $M(A)$ is a solution of the following equation [2, 12, 35].

$$(\epsilon A + a A^3 + \frac{h^2}{2} A)\frac{dM}{dA} + \frac{h^2}{2} A^2 \frac{d^2 M}{dA^2} = -1 \tag{30}$$

At $A = R$ we have an absorbing boundary with $M(A) = 0$ and at $A = 0$ a "built-in" boundary which can be treated via a proper limit process. In our case we have to take a second absorbing boundary at δ and have to consider the limit $\delta \rightarrow 0$ [28]. Then we obtain

$$M(A) = \frac{2}{h^2} \int_A^R v^{-\frac{2\epsilon}{h^2}-1} \exp(-\frac{a}{h^2} v^2) \int_0^v u^{\frac{2\epsilon}{h^2}-1} \exp(\frac{a}{h^2} u^2)\, du\, dv \tag{31}$$

$M(A)$ is the mean time of a realization of process (4) to reach R the first time starting at A. Thus, as a first step we show the identity $M(A) \equiv T_A$. For this purpose the corresponding integrals are splitted into two parts :

$$M(A) = \frac{2}{h^2} \int_A^R v^{-\frac{2\epsilon}{h^2}-1} \exp(-\frac{a}{h^2} v^2)\, dv \int_0^A u^{\frac{2\epsilon}{h^2}-1} \exp(\frac{a}{h^2} v^2)\, du$$

$$+ \frac{2}{h^2} \int_A^R \int_A^v v^{-\frac{2\epsilon}{h^2}-1} u^{\frac{2\epsilon}{h^2}-1} \exp(\frac{a}{h^2}[u^2 - v^2])\, du\, dv \tag{32}$$

$$T_A = \frac{2}{h^2} \int_0^A v^{\frac{2\epsilon}{h^2}-1} \exp(\frac{a}{h^2} v^2)\, dv \int_A^R u^{-\frac{2\epsilon}{h^2}-1} \exp(-\frac{a}{h^2} u^2)\, du$$

$$+ \frac{2}{h^2} \int_A^R \int_v^R v^{\frac{2\epsilon}{h^2}-1} u^{-\frac{2\epsilon}{h^2}-1} \exp(\frac{a}{h^2}[v^2 - u^2])\, du\, dv \tag{33}$$

The equivalence of the first terms of the sums is obvious since we have in both cases the same product of two one-dimensional integrals. The identity of the second terms can be proven by appropriate substitution. In fact, after an exchange of u and v we have the same integrand and integrate over the same region.

Now it has to be shown that the average over $M(A)$ (or T_A) equals $\bar{T}_B$. Again we split the corresponding integral (19):

$$
\bar{T}_B = \frac{2}{h^2} \int_0^B v^{\frac{2\zeta}{h^2}-1} \exp(\frac{a}{h^2}v^2)
$$

$$
\left[\frac{1}{B} \int_v^B u^{-\frac{2\zeta}{h^2}} \exp(-\frac{a}{h^2}u^2)\,du + \int_B^R u^{-\frac{2\zeta}{h^2}-1} \exp(-\frac{a}{h^2}u^2)\,du \right] \qquad (34)
$$

$$
+ \frac{2}{h^2} \int_B^R v^{\frac{2\zeta}{h^2}-1} \exp(\frac{a}{h^2}v^2) \int_v^R u^{-\frac{2\zeta}{h^2}-1} \exp(-\frac{a}{h^2}u^2)\,du\,dv
$$

A comparison with Eq. 33 leads to

$$
\bar{T}_B = M(B) + \frac{2}{Bh^2} \int_0^B v^{\frac{2\zeta}{h^2}-1} \exp(\frac{a}{h^2}v^2) \int_v^B u^{-\frac{2\zeta}{h^2}} \exp(-\frac{a}{h^2}u^2)\,du\,dv \qquad (35)
$$

Now we integrate the averaged time $M(A)$ by parts:

$$
\frac{1}{B} \int_0^B M(A)\,dA = M(B) - \frac{1}{B} \int_0^B A\frac{dM}{dA}\,dA \qquad (36)
$$

From Eq. 32 we derive easily

$$
\frac{dM}{dA} = -\frac{2}{h^2} A^{-\frac{2\zeta}{h^2}-1} \exp(-\frac{a}{h^2}A^2) \int_0^A u^{\frac{2\zeta}{h^2}-1} \exp(\frac{a}{h^2}u^2)\,du\,dA \qquad (37)
$$

Inserting this expression into formula (36) the equivalence to $\bar{T}_B$ can be seen using appropriate changes of variables. In this way we have exemplified the identity of the different approaches and our somewhat unusual but very convenient calculation of the mean length of laminar phases is justified.

Acknowledgements

One of us (H.H.) is very grateful to R.L. Stratonovich, Yu.L. Klimontovich, P.S. Landa und L. Schimansky-Geier for helpful discussions.

References

1. Y. Pomeau, P. Manneville: Commun. Math. Phys. 74(1980)189.
2. J.P. Eckmannn, L. Thomas, P. Wittwer: J. Phys. A 14(1981)3153.
3. J.E. Hirsch, B.A. Huberman, D.J. Scalapino: Phys. Rev. A 25(1982)519.
4. G. Mayer-Kress, H. Haken: Phys. Lett. A 82(1981)151.

5. J. Maurer, A. Libchaber: J. Physique Lett. 41(1980)L-515.
6. M. Dubois, M.A. Rubio, P. Bergé: Phys. Rev. Lett. 51(1983)1446.
7. J.Y. Huang: Phys. Rev. A 36(1987)1495.
8. R. Richter, U. Rau, A. Kittel, G. Heinz, J. Parisi, R. P. Huebener: Z. Naturfors ch. 46a(1991)1012.
9. N. Kreisberg, W.D. McCormick, H.L. Swinney, Physica D 50(1991)463.
10. H. Herzel, P. Plath, P. Svensson: Physica D 48(1991)340.
11. E. Ringuet, C. Roze, G. Gouesbet: Phys. Rev. E 47(1993)1405.
12. R.L. Stratonovich: *Topics in the Theory of Random Noise.* Gordon and Breach, New York, 1963.
13. G. Nicolis, I. Prigogine: *Self-Organization in Non-Equilibrium Systems.* John W ley and Sons, New York, 1977
14. A. Schenzle, H. Brand: Phys. Rev. A 20(1979)1628.
15. W. Ebeling, R. Feistel: *Physik der Selbstorganisation und Evolution.* Akademie-Verlag, Berlin 1982.
16. W. Horsthemke, R. Lefever: *Noise-Induced Transitions.* Springer-Verlag, Berlin 1982.
17. P. Richetti, F. Argoul, A. Arneodo: Phys. Rev. A 34(1986)726.
18. F. Argoul, A. Arneodo, P. Richetti: J. Physique 49(1988)767.
19. H. Herzel, B. Pompe: Phys. Lett. A 122(1987)121.
20. H. Herzel: Z. angew. Math. Mech. 68(1988)582.
21. P. Berge, Y. Pomeau, Ch. Vidal: *Order within Chaos.* John Wiley and Sons, New York 1986.
22. F. Argoul, A. Arneodo: J. Phys. France Lett. 46(1985)L-901.
23. F. Argoul, A. Arneodo: Lect. Notes Math. 1186(1986)338.
24. P.S. Landa, R.L. Stratonovich: Radiophys. Quantum Electron. (USA) 30(1987)53.
25. A.S. Pikovsky: J. Phys. A 16(1983)L-109.
26. Yu.L. Klimontovich, M. Bonitz: Pisma Sh. TF 12(1986)1349.
27. P.S. Landa: Vestnik Mosc. Univ. 28(1987)22.
28. N.S. Goel, N. Richter-Dyn: *Stochastic Models in Biology.* Academic Press, New York, 1974.
29. Yu.L. Klimontovich: *Kinetic Theory of Electromagnetic Processes.* Springer-Verlag, Berlin 1983.
30. W. Ebeling, H. Engel-Herbert: Physica A 104(1980)378.
31. H. Malchow, L. Schimansky-Geier: *Noise and Diffusion in Bistable Nonequilib -ium Systems.* Teubner-Verlag Leipzig, 1985.
32. Yu.I. Frenkel: *Kinetic Theory of Liquid (in russ.).* Leningrad, 1945.
33. F.T. Arecchi, A. Politi: Phys. Rev. Lett. 45(1980)1219.
34. P. Haenggi: Phys. Lett. A 78(1980)304.
35. K. Lindenberg, K.E. Shuler, J. Freeman, T.J. Lie: J. Stat. Phys. 12(1975)217.

ASPECTS OF LIOUVILLE INTEGRABILITY
IN QUANTUM MECHANICS

J. HIETARINTA*
LPN, Université Paris VII, Tour 24-14, 5ème étage
2 Place Jussieu, 75251 Paris, France

Abstract. When the classical concept of Liouville integrability (i.e the existence of commuting constants of motion) is extended to quantum mechanics there are new purely quantal problems related to the number and independence of the commuting quantities. We discuss these problems with illustrations from one- and two-dimensional Hamiltonian systems.

1. Introduction

In quantum mechanics a good definition for chaos is still missing. In classical mechanics the situation is much better: there are several more or less equivalent criteria for chaos. For example we can focus on the sensitive dependence on initial conditions in a bounded system: if phase space trajectories starting from nearby points diverge exponentially then we may call the system chaotic. Clearly this particular definition cannot be straightforwardly extended to quantum mechanics, because due to the uncertainty relations it is impossible even in principle to prepare a quantum system for a precisely defined point in phase space. (In [1] a quantum "sensitivity operator" was defined and then shown to have no positive Liapunov exponents.) Thus in general quantum effects seem to smooth out the very fine scale structure, this is often called "quantum suppression of chaos". (It has been argued, for example, that in quantum mechanics the perturbative small denominator problem disappears [2].)

Although the above attempt to define quantum chaos fails there may be others that work. One should therefore study also the other definitions of classical chaos and check if any one of them can be extended into quantum mechanics somehow. In this talk we consider "Liouville integrability". For a classical Hamiltonian system Liouville integrability is the opposite of chaos, hence if we can define a quantum equivalent integrability then we can perhaps have a provisional definition of quantum chaos by negation.

2. The definition of Liouville integrability

We will start with the following
  Definition: A N-dimensional Hamiltonian system is called *Liouville integrable* if the following conditions hold:
    1. There are N functions I_i of the N coordinates and N momenta (one of them, say I_1 is the Hamiltonian) such that

---

* On leave of absence from: Department of Physics, University of Turku, 20500 Turku, Finland

Z. Haba et al. (eds.), Stochasticity and Quantum Chaos, 115–121.
© 1995 *Kluwer Academic Publishers*.

2. the I_i are *in involution* i.e. $\{I_i, I_j\} = 0$,

3. the I_i are *independent*,

4. the I_i are *nice* functions.

For this to be a good definition the italiced qualifications must be made precise, and they do indeed have different meaning and importance in classical and quantum mechanics.

2.1. LIOUVILLE INTEGRABILITY IN CLASSICAL MECHANICS

When the above definition is used in classical mechanics, involution is defined by the *Poisson bracket*, independence means *functional* independence, and *analytic* functions are certainly nice enough. If a system is Liouville integrable in this sense, then it has very good properties [3], e.g.:

— The phase space is foliated into smooth manifolds on which the motion stays. In compact systems the foliation is into tori.

— In principle one can solve the system in quadratures (Liouville's theorem).

We should also note the following special situations:

— If the I_i are all quadratic in momenta, then one can separate variables in some coordinate system.

— If one finds more than N independent functions that commute with the Hamiltonian (but then necessarily some pairs will not be in involution) one can get further simplifications (superintegrability or algebraic integrability).

2.2. LIOUVILLE INTEGRABILITY IN QUANTUM MECHANICS

When one tries to define Liouville integrability in quantum mechanics it turns out that only part 2) of the definition above extends easily: In quantum mechanics the invariants are in involution if they commute as operators. [It is then assumed that the basic quantities p and q are quantized so that $[\hat{q}_i, \hat{p}_j] = i\hbar\delta_{ij}$.] The other parts of the above definition need careful analysis because of the following new aspects:

1. It turns out to be meaningful and useful to have more than N commuting quantities (of course they cannot all be functionally independent).

3. Commuting operators can never be completely independent, but can be written as functions of one single operator (von Neumann). (The implications of this result on quantum integrability were discussed in [4])

We will discuss these problems more in the following section, but let us first see how one can make (at least formally) the transition from classical to quantum mechanics while preserving integrability. The problem is the following: Suppose we have N functions of p and q in involution, how can we construct from them N operators that commute?

The standard way of constructing quantum operators from classical quantities is by replacing p_k by $-i\hbar\partial_{q_k}$. This may introduce ordering ambiguities in monomials which for some k contain $p_k^n q_k^m$. Various ordering rules have been introduced to guide in this, for example according to the Weyl rule

$$p^n f(q) \rightarrow \frac{1}{2^n} \sum_{l=0}^{n} \binom{n}{l} \hat{p}^{n-l} f(\hat{q}) \hat{p}^l$$

However, it turns out that there is no one rule that is best for conserving integrability.

For two dimensional systems with Hamiltonians of type $H = \frac{1}{2}(p_x^2 + p_y^2) + V(x, y)$ (in which case integrable systems should have two commuting quantities, the Hamiltonian and a second invariant) the above question was studied at length in [5]. The results can be summarized as follows:

- If the invariants are quadratic then quantization with the Weyl rule works.
- Sometimes this works for higher order (in momenta) invariants as well.
- In many cases one needs to modify a higher order invariant with $\hbar^2$ terms.
- In some cases one has to modify the Hamiltonian as well.

The last result is surprising since the Hamiltonian has no ordering ambiguity. An illustration of this phenomena is given by the Fokas-Lagestrom Hamiltonian

$$H = \frac{1}{2}p_x^2 + \frac{1}{2}p_y^2 + (xy)^{-2/3} - \hbar^2 \frac{5}{72}\left(\frac{1}{x^2} + \frac{1}{y^2}\right), \tag{1}$$

$$I = p_x p_y (x p_y - y p_x) + 2(xy)^{-2/3}(x p_x - y p_y) - \hbar^2 \frac{5}{36}\left(\frac{x p_x}{y^2} - \frac{y p_y}{x^2}\right). \tag{2}$$

The terms without the $\hbar^2$ coefficient constitute the classical Hamiltonian and its second invariant. If we construct quantum operators from (1,2) without the $\hbar^2$-terms (with any ordering rule) it turns out that the resulting operators do *not* commute. However, with the above deformations a commuting pair of operators is obtained with the Weyl rule[6]. Note that the deformations cannot be explained by any ordering rule, because there is no ordering ambiguity in the Hamiltonian.

The reason for the extra $\hbar^2$ terms is the following: There does exist a certain classically integrable Hamiltonian H' with a second invariant I', which *can* be quantized straightforwardly by the Weyl rule while preserving integrability; the above system (1,2) is obtained from H', I' by a canonical transformation followed by a time change (or 'coupling constant metamorphosis'). It turns out that these transformations generate the extra terms [7], and this seems to be the general rule with the deformation terms [8].

3. One dimensional systems with two commuting differential operators

Let us now return to discuss the open questions in the definition of quantum Liouville integrability. Let us consider the simplest case, a one dimensional Hamiltonian system. In classical mechanics it is trivially integrable since one needs only a single invariant which is provided by the Hamiltonian itself. Indeed the system can be solved by quadratures in the standard way.

When we consider the corresponding quantum system, it is, by definition, Liouville integrable as well, but this knowledge does not help us at all in solving e.g. for the wave function and the spectrum. (One may of course take the attitude that any linear equation is called solvable.)

The surprising observation is that for some potential we can have two commuting operators H and I (which must of course be algebraically related, but not in the simple way that one is a function of the other). If we just consider the leading (in derivatives) part of I it is easy to see that is must be a pure power of ∂_x. Furthermore any even leading power can be eliminated by subtracting a suitable power of H. Thus the simplest nontrivial commuting pair is obtained with the ansatz

$$I_2 = -i\hbar^3 \partial_x^3 + \frac{1}{2}(\partial_x C(x) + C(x) \partial_x).\tag{3}$$

If one calculates the commutator of H and I_2, it vanishes if [9]

$$V(x) = \hbar^2 \mathcal{P}(x), \quad C(x) = 3\hbar^2 \mathcal{P}(x),\tag{4}$$

where $\mathcal{P}$ is the Weierstrass elliptic function. Naturally H and I_2 are algebraically related, indeed one finds that $f(H, I_2) = 0$, where

$$f(x, y) := 8x^3 - y^2 - \frac{1}{2}\hbar^4 x g_2 + \frac{1}{4}\hbar^6 g_3\tag{5}$$

and the g_i are the parameters that appear in the definition of $\mathcal{P}$ [10].

What can we then do with this information? By playing H and I_2 against each other one can derive a first order equation for the common eigenfunction and therefore express it in quadratures, the result is [9]

$$\Psi = (\mathcal{P}(x) - \mathcal{P}(\alpha))^{1/2} \exp\left(\pm\frac{1}{2}\mathcal{P}'(\alpha)\int^{\mathcal{P}(x)}(4t^3 - g_2 t - g_3)^{-1/2}dt\right)\tag{6}$$

where α is related to the energy by $E = -\frac{1}{2}\hbar^2 \mathcal{P}(\alpha)$. Of course one still has to find which values of α give squared integrable eigenfunctions, but having an expression for Ψ makes this problem much easier.

The conclusion we make about the 'independence' question above is that even *algebraically* dependent commuting differential operators are useful for solving the system, as long as no one of them can be expressed as a function of the others. Thus in quantum mechanics the requirements of independence may be relaxed significantly.

In fact the problem of finding commuting sets of differential operators is quite old, it was analysed already in the 1920's by Burchnal and Chaundy [11]. More recently the problem was studied in the context of soliton theory by Dubrovin, Krichever and Novikov in the 1970's [12]. They found that the spectra of systems with a commuting pair of differential operators have nice properties. This is related to the existence of a Lax pair for a time independent potential (=static soliton). These results are now being extended to higher dimensions [13].

4. Problems with the number of commuting operators

Let us next look at the question of dependence from a completely different point of view. A theorem of von Neumann [14] says that "If A, B are two commuting Hermitean operators, then there exists a Hermitean operator R, of which both are functions, i.e., $A = r(R)$, $B = s(R)$." The conclusion from this is that commuting operators are never truly independent: one can give all of them in terms of a single one. Thus the number of commuting operators seems to be ambiguous.

The above theorem is nothing but a consequence of the familiar fact that commuting operators have simultaneous eigenfunctions. The countable set of simultaneous eigenfunctions can be mapped to the positive integers and then, using projection operators, one can do the reverse and construct commuting sets of operators with these same eigenfunctions.

4.1. THE NUMBER OF OBSERVATIONS NEEDED FOR A MAXIMAL OBSERVATION

At this point we should note that in many books of quantum mechanics one talks about a complete set of commuting operators and often emphasis is placed on the cardinality of this set. But from the theorem above we must conclude that [15] "The number of operators in such a complete set is *not* a characteristic of the dynamical system under investigation".

Let us illustrate the above two observations with a particle moving in the spherically symmetric Coulomb potential $V(r) = -Ze^2/r^2$ [16][17]. It is well known that the Hamiltonian H, total angular momentum L^2 and the z-component L_z of the angular momentum are the three commuting operators relevant to this system. If one now makes a measurement of H one finds that its eigenvalue E can be written as $E = -RchZ^2/n^2$ for some integer n. This does not determine the state uniquely (unless $n = 1$), but the state can be shown to be n^2 fold degenerate. In order to further determine the state one makes a measurement on L^2, and finds that its eigenvalue M^2 can be written as $M^2 = \hbar^2 l(l+1)$, for some integer $0 \leq l \leq n-1$. The state is still not determined completely (unless $l = 0$) but is $2l + 1$ fold degenerate. Thus a third measurement is needed, that of L_z, its eigenvalue must be $M_z = \hbar m$ for some integer $-l < m < l$.

Thus in this example it seems that we need *three* measurements to fully specify the state, i.e. to make a "maximal or complete observation"[16]. But since we know what the possible eigenstates of these operators are, it is enough to make just one observation: In this case we can take, for example, $RchZ^2/H + \sqrt{2}L^2\hbar^{-2} + \sqrt{3}\hbar L_z\hbar^{-1}$ as the single operator whose eigenvalues can be used to determine the state completely: The measurement of this operator will yield an eigenvalue, which can be written uniquely as $n^2 + \sqrt{2}l + \sqrt{3}m$, where n, l, m are integers mentioned above.(Of course this is not useful in practice, especially for high quantum numbers.) Thus only one observation is needed, if the operator is chosen suitably.

4.2. FRACTAL RELATIONS

Let us consider in more detail the commuting operators L^2 and L_z. Let us define a new operator by $R := L^2 + \hbar L_z$. From the spectra of L^2 and L_z given above it is easy to see that the spectrum of R is $\hbar^2 n$, where $n = 0, 1, 2, \ldots$ and each common eigenstate of L^2 and L_z is uniquely characterized by the eigenvalue of R. What has happened is that we have mapped the discrete eigenvalues of L^2 and L_z, which span a two dimensional lattice, to the set of nonnegative integers. In order to complete the statement given in the theorem of von Neumann mentioned above, we have to express L^2 and L_z in terms of R, i.e. we need a map from integers back to the two dimensional lattice. This is provided by the following:

$$L^2 = \hbar^2[\hbar^{-1}\sqrt{R}]([\hbar^{-2}\sqrt{R}] + 1), \quad L_z = \hbar^{-1}(R - L^2).$$

Here [.] stands for the integer part of the quantity (this function can be smoothed to make the operation analytic). It is clear that these functions reproduce the eigenvalues of L^2 and L_z.

Before commenting on the above constructions let us recall how fractals can be used to identify chaos in classical mechanics. As was noted at the beginning, the existence of a sufficient number of commuting quantities implies a smooth foliation of the phase space (into tori, if the system is bounded). In other words, an invariant provides a map from the phase space to the reals, such that the map commutes with the time evolution given by the Hamiltonian. But such a map can be constructed for any dynamical system as follows: Take any initial value in phase space and integrate the motion forward to $t = \infty$ and backward to $t = -\infty$. This defines a curve in phase space; we ignore its parametrization by time. To this curve we associate the intial value (which is a vector), and if we combine a family of initial values we get invariant manifolds of any given dimension. Of course this construction is useless, because there is no guarantee of smoothness of this foliation. Indeed, we know now that chaotic systems can have tangles around some orbits, and at these points the supposed foliation of the phase space will exhibit fractal behavior (i.e. fractal dimension > topological dimension). Thus we use fractality to separate good foliations from bad ones, and in the latter case the system is said to be chaotic.

Let us now return to the problem of number of invariants in quantum mechanics. The point is that in the above construction the resulting function pair (r, s) is (in the $\hbar \to 0$ limit) a fractal function mapping the positive real line to the sector $\{(x, y)|x > 0,\ x < y < -x\}$. In classical mechanics we use the fractality as a test of chaos in the sense that fractal foliations imply chaos. Thus when we discuss integrability vs. chaos in classical mechanics we do not allow functions that change the fractal dimension of the object. The same principle should be used in quantum mechanics. Of course fractality only appears in the classical limit, because the quantum spectrum is discrete.

5. Conclusions

In this talk we have shown that the basic contents of Liouville integrability is the same in classical and quantum mechanics. However, there are important differences related to independence and number of the commuting quantities. Quantum mechanics turns out to be richer (and more interesting) in this respect.

The first observation is that it makes sense and is useful to have even algebraically dependent commuting (differential) operators, it is just sufficient that no operator is a function of the others.

Secondly, the dependence that is used to show that all commuting operators are functions of a single operator is really of fractal type, and should not be allowed.

One of the open questions that needs further work is find a proper class of operators that one should consider in discussing chaos v.s. integrability. If we work only within the ring of differential operators the above problems with independence are avoided. Certainly one can extend the class of operators further, but the characterization on the best class is still open.

Acknowledgements

I would like to thank B. Grammaticos for comments of the manuscript.

References

1. M.H. Partovi, Phys. Rev. A **45**, 555 (1992).
2. M. Robnik, J. Phys. A: Math. Gen. **19**, L841 (1986).
3. V.I. Arnold, *Mathematical Methods of Classical Mechanics*, (Springer, 1978) p. 271.
4. S. Weigert, Physica D **56**, 107 (1992).
5. J. Hietarinta, J. Math. Phys. **25**, 1833 (1984).
6. J. Hietarinta, B. Grammaticos, B. Dorizzi and A. Ramani, C. R. Acad. Sci. Paris Serie II, **297**, 791 (1983).
7. J. Hietarinta, B. Grammaticos, B. Dorizzi, and A. Ramani, Phys. Rev. Lett. **53**, 1707 (1984).
8. J. Hietarinta and B. Grammaticos, J. Phys. A: Math. Gen. J. Phys. A: Math. Gen. **22**, 1315 (1989).
9. J. Hietarinta, J. Phys. A: Math. Gen. **22**, L143 (1989).
10. M. Abramowitz and I.S. Stegun, *Handbook of Mathematical Functions*, (Dover, 1965), p. 629.
11. J.L Burchnal and T.W. Chaundy, Proc. Lond. Math. Soc. 21, 420 (1922), Proc. Royal Soc. (London) **118**, 557 (1928).
12. B.A. Dubrovin, I.M. Krichever and S.P. Novikov, Sov. Math. Dokl. **17**, 947 (1976).
13. O.A. Chaklyh and A.P. Veselov, Commun. Math. Phys. **126**, 597 (1990).
14. J. von Neumann, *Mathematical Foundations of Quantum Mechanics*, (Princeton UP, 1955) p. 173.
15. E.C. Kemble, *The Fundamental Principles of Quantum Mechanics*, (Dover, 1958) p. 287.
16. F. Mandl, *Quantum Mechanics*, 2nd ed. (Butterworths, 1957) p. 79-80.
17. E. Ikenberry, *Quantum Mechanics for Mathematicians and Physicists*, (Oxford UP, 1962) p. 159.

KAM TECHNIQUES FOR TIME DEPENDENT QUANTUM SYSTEMS

H.R. JAUSLIN
Laboratoire de Physique de l'Université de Bourgogne
6, bd. Gabriel; 21100 Dijon, France.
e-mail: jauslin@satie.u-bourgogne.fr

Abstract. We consider a spin 1/2 in constant magnetic field perturbed by a quasiperiodic time dependent magnetic field. We discuss its stability properties in terms of the spectrum of the corresponding quasienergy operator. Since the spectrum of the unperturbed problem is dense, there appear small denominators in the perturbation theory, corresponding to resonances. They are treated with a technique developped by L.H. Eliasson, based on a KAM iteration.

1. Introduction

The understanding of quantum systems subjected to time dependent forces has advanced considerably [1] - [3]. The case of periodic fields has been studied in great detail. It was noticed early that quantum systems tend to be more stable than their classical counterparts. Models like the periodically kicked rotator, which classically show an unbounded diffusive growth of the energy associated to chaotic trajectories, show a saturation of the energy growth in its quantum version. Later studies indicated that this quantum limitation of diffusion is weaker or even absent for nonperiodic perturbations [4]. Here we are going to discuss the case of quasiperiodic perturbations of systems evolving in a finite dimensional Hilbert space. It is known [5][6] that such systems can already display quite complicated evolutions in Hilbert space. We are going to discuss their stability properties under small perturbations. We consider a spin 1/2 in a time dependent magnetic field described by a Hamiltonian acting on the Hilbert space $\mathcal{H} = \mathbf{C}^2$ of the form

$$H = H_0 + \varepsilon V(\underline{\omega} t + \underline{\theta})$$

where

$$H_0 = \beta_0 \begin{pmatrix} 1 & 0 \\ 0 & -1 \end{pmatrix}$$

and V is a hermitian 2×2 matrix that we can take of trace zero, without restriction of generality.

We consider a quasiperiodic perturbation with two incommensurate frequencies, $\underline{w} = (\omega_1, \omega_2)$, and initial phases $\underline{\theta} = (\theta_1, \theta_2)$; V is periodic in each variable: $V(\underline{\theta} + \underline{n}2\pi) = V(\underline{\theta})$, for any $\underline{n} = (n_1, n_2) \in \mathbf{Z}^2$.

We want to discuss the following question of stability: The solution of the unperturbed equation

$$i\frac{\partial \psi^0}{\partial t} = H_0 \psi^0$$

123

Z. Haba et al. (eds.), Stochasticity and Quantum Chaos, 123–129.

can be expanded in terms of the eigenfunctions defined by $H_0\varphi_m = E_m\varphi_m$ as

$$\psi^0(t) = \sum_m c_m e^{-iE_m t}\varphi_m; \qquad c_m = \langle \varphi_m, \psi^0(t=0)\rangle$$

The function $\psi^0(t)$ is therefore almost-periodic. We will say that the system is stable if the solution $\psi(t)$ of the perturbed equation

$$i\frac{\partial\psi}{\partial t} = H(t)\psi \tag{1.1}$$

is still almost-periodic. This question can be studied by analyzing the spectral properties of the quasienergy operator

$$K = -i\underline{w}\frac{\partial}{\partial\underline{\theta}} + H(\underline{\theta}), \qquad H(\underline{\theta}) = H_0 + \varepsilon V(\underline{\theta})$$

defined on an enlarged Hilbert space $\mathcal{K} = \mathcal{H} \otimes L_2(\mathbf{T}^2, d\theta_1 d\theta_2)$, where $\mathbf{T}^2$ is a two dimensional torus [7]. The relation between its spectrum and the dynamics follows from the relation [7][6]

$$e^{-iKt} = \mathcal{T}_{-t}U(t,0 \ ; \ \underline{\theta})$$

where $U(t,0 \ ; \ \underline{\theta})$ is the propagator of the equation 1.1 and $[\mathcal{T}_{-t}\psi](\underline{\theta}) = \psi(\underline{\theta} - \underline{w}t)$. If the spectrum of the quasienergy operator K is pure point with eigenfunctions $\psi_\nu(\underline{\theta})$ satisfying $K\psi_\nu = \lambda_\nu \psi_\nu$ then the time evolution of an initial condition $\varphi(0)$ can be written as [6]

$$\varphi(t) = \sum_\nu c_\nu e^{-i\lambda t}\psi_\nu(\underline{\theta} + \underline{w}t), \qquad c_\nu = <\varphi(0)\otimes 1 \ , \ \psi_\nu >_\mathcal{K}$$

Here the scalar product is the one of the enlarged space $\mathcal{K}$. This provides a generalization of the eigenfunction expansion to time dependent problems which implies in turn that the evolution is almost-periodic. A practical interpretation of this fact is that the dynamics can be described essentially in a finite dimensional subspace of $\mathcal{K}$. The error made by truncating to a finite dimensional subspace is uniformly bounded in time. This property becomes quite useful for the numerical treatment of these problems [8].

It is however not trivial to control time-dependent perturbations. The reason can be explained intuitively as follows. We can consider an unperturbed quasienergy operator $K_0 = -i\underline{w}\partial/\partial\underline{\theta} + H_0$. Its eigenfunctions and eigenvalues can be written explicitly as

$$\psi^0_{m,\underline{n}} = e^{-i\underline{n}\underline{\theta}}\varphi_m, \qquad \lambda^0_{m,\underline{n}} = E_m + n_1\omega_1 + n_2\omega_2, \qquad \underline{n} = (n_1, n_2), \quad m_j \in \{+1, -1\}$$

$$\varphi_+ = \begin{pmatrix} 1 \\ 0 \end{pmatrix}, \quad \varphi_- = \begin{pmatrix} 0 \\ 1 \end{pmatrix}, \qquad E_\pm = \pm\beta_0$$

Since ω_1/ω_2 is irrational, the spectrum of the unperturbed K_0 is pure point but dense on the real line. As a consequence one cannot apply the usual analytic perturbation theory [9]. Indeed the expression for the eigenvalues would be of the form

$$\lambda^\varepsilon_{m,\underline{n}} = \lambda^0_{m,\underline{n}} + \varepsilon V_{m,\underline{n};m,\underline{n}} + \varepsilon^2 \sum_{m',\underline{n}'} \frac{|V_{m,\underline{n},m',\underline{n}'}|^2}{\lambda^0_{m,\underline{n}} - \lambda^0_{m',\underline{n}'}} + O(\varepsilon^3)$$

where

$$V_{m,\underline{n};m',\underline{n}'} = <\psi^0_{m,\underline{n}} , V \psi^0_{m',\underline{n}'} >_\mathcal{K}$$

Since the unperturbed spectrum is dense, the denominator $\lambda^0_{m,\underline{n}} - \lambda^0_{m',\underline{n}'} = E_m - E_{m'} + \underline{w}(\underline{n} - \underline{n}')$ becomes arbitrarily small for infinitely many terms in the sum, which breaks its convergence. In order to treat this kind of problem one has to use different methods, based on KAM techniques that were originally develcped to deal with similar small denominator problems in classical mechanics.

2. KAM techniques

In order to prove that the perturbed quasienergy operator has point spectrum the goal is to construct a unitary transformation R such that

$$R^\dagger K R = -i\underline{w}\frac{\partial}{\partial \underline{\theta}} + \beta_\infty \begin{pmatrix} 1 & 0 \\ 0 & -1 \end{pmatrix} \equiv D_\infty$$

The transformation R is constructed by composition of a sequence of operators constructed by iteration

$$R = \ldots R_3 R_2 R_1 \tag{2.1}$$

To describe how the R_j are constructed we discuss R_1. The perturbed quasienergy operator K can be separated into two parts

$$K \equiv K_1 = D_1 + V_1$$

where $D_1 \equiv K_0$ is diagonal in the basis of eigenfunctions of K_0 and $V_1 \equiv \varepsilon V$ is nondiagonal and of order $O(\varepsilon)$. The general idea is that R_1 is to be constructed such that

$$R_1^\dagger(D_1 + V_1)R_1 = D_2 + V_2$$

with $D_2 = -i\underline{w}\partial/\partial\underline{\theta} + \beta_2 \begin{pmatrix} 1 & 0 \\ 0 & -1 \end{pmatrix}$ diagonal in the unperturbed eigenbasis, and V_2 a nondiagonal rest of order $O(\varepsilon^2)$. By iterating this procedure the nondiagonal part is reduced successively to $O(\varepsilon^4)$, $O(\varepsilon^8)\ldots$, and it vanishes in the limit of infinite iterations.

We will use the following notation for the matrix elements of e.g. V:

$$\tilde{V}(m, m'; \underline{n}, \underline{n}') \equiv \tilde{V}(m, m'; \underline{n} - \underline{n}') = \left\langle \varphi_m \left| \int d\underline{\theta}\, e^{i(\underline{n}-\underline{n}')\underline{\theta}} V(\underline{\theta}) \right| \varphi_{m'} \right\rangle$$

i.e. $V(m, m'; \underline{k})$ is the $\underline{k}$-th term of the Fourier series of V. R_1 being unitary, it can be represented as $R_1 = e^{-W_1}$, with $W_1^\dagger = -W_1$. Using the Taylor expansion of the exponential we can write

$$e^W (D_1 + V_1)e^{-W} = D_1 + V_1 + [W_1, D_1] + V_2$$

with

$$V_2 = [W_1 , ([W_1, D_1]/2 + V_1)] + O(\varepsilon^3)$$

The condition that all nondiagonal terms of order $O(\varepsilon)$ disappear is expressed as

$$V_1 + [W_1, D_1] = \delta\beta_1 \begin{pmatrix} 1 & 0 \\ 0 & -1 \end{pmatrix} \qquad (2.2)$$

i.e. we have to determine W_1 and the correction to the diagonal part $\delta\beta_1 \equiv \beta_2 - \beta_1$. Equation 2.2 can be solved by expressing it in terms of matrix elements with respect to the unperturbed basis (which amounts to work in Fourier space). The diagonal elements of the perturbation determine $\delta\beta_1$:

$$\delta\beta_1 = \tilde{V}_1(m = +1, m = +1; \underline{n} = 0)$$

and

$$\tilde{W}_1(m, m'; \underline{n}) = \frac{\tilde{V}_1(m, m'; \underline{n})}{\underline{w}\underline{n} + \beta_1(m - m')} \qquad (2.3)$$

The technique used in earlier work [2][10][6] was to treat at each iteration only the sets of parameters $\underline{w}, \beta_j$ that satisfy a condition of diophantine type

$$\left| \underline{w}\underline{n} + (m - m')\beta_j \right| > \frac{\gamma_j}{|\underline{n}|^\tau} \qquad (2.4)$$

which suffices to guarantee the convergence of the Fourier series of W_j, as well as the convergence of the iteration and of the composition 2.1. If V is analytic, its Fourier coefficients decrease exponentially, and compensate the growth $\sim |\underline{n}|^\tau$ produced by the denominator. The parameters that don't satisfy this property are discarded. One proves that for small perturbations the measure of this set is small. Since a condition 2.4 has to be imposed at each step of the iteration and β_j is different at each step, one completely loses the information on which parameters have actually been retained. An improvement of this technique has been developed by L.H. Eliasson [11][12], which allows to obtain stronger and more precise results. The main idea is not to discard the case of resonance, where the diophantine condition 2.4 is not satisfied, but to treat it with a different transformation S_j. The main aspects of this technique can be sketched as follows. The goal is to obtain the following sequence of bounds

$$\|V_j\| < \varepsilon_j, \quad \text{with } \varepsilon_{j+1} = \varepsilon_j^{1+\sigma}, \quad \sigma \in (0,1); \qquad \text{i.e. } \varepsilon_j = \varepsilon_1^{(1+\sigma)^j} \xrightarrow{j \to \infty} 0$$

One makes throughout the assumption that $\underline{w}$ is diophantine, i.e. there are constants $c_0 > 0, \tau > 2$ such that

$$|\underline{w}\underline{n}| > |c_0|/|\underline{n}|^\tau, \forall \underline{n} \in \mathbf{Z}^2. \qquad (2.5)$$

We remark that the difference with the condition 2.4 is that β does not intervene, and the condition stays unchanged for all iterations. The set of $\underline{w}$'s satisfying this condition is of full measure.

One starts by introducing at each iteration an "ultraviolet cutoff"

$$V_j = V_j^{\leq N_j} + V_j^{> N_j}$$

where $V_j^{< N_j}$ contains the Fourier terms up to $|\underline{n}| \leq N_j$. The cutoff N_j is chosen such that $\|V_j^{> N_j}\| < \varepsilon_{j+1}$, while by construction $\|V_j^{\leq N_j}\| < \varepsilon_j$. Since the higher

Fourier terms $V_j^{>N_j}$ already satisfied the required bound they can be left unchanged and the transformation R_j is constructed to take care only of the lower terms $V_j^{\leq N_j}$. One easily verifies that one can take

$$N_j \sim \ln 1/\varepsilon_j.$$

This logarithmic dependence plays a crucial role at several points of this method.

One has then to distinguish two cases: We say that there is an (ε_j, N_j)-resonance if there is one (or several) $\underline{n}$ with $|\underline{n} \leq N_j$ such that $|\underline{nw} - 2\beta| < \varepsilon_j$. The (ε_j, N_j) non-resonant case can be treated with the usual method, i.e. one determines the transformation R_j from 2.3.

For the resonant case we construct a different transformation. One first remarks that the choice $N_j \sim \ln 1/\varepsilon_j$ implies that there is at most one $\underline{n}_j^*$ that produces (ε_j, N_j)-resonance. The new transformation can therefore be adapted to handle this single resonance. In a first step we apply a transformation

$$S_j = e^{i\underline{n}_j^* \theta/2} \begin{pmatrix} 1 & 0 \\ 0 & -1 \end{pmatrix}$$

which eliminates the resonance:

$$S_j^{-1}(D_j + V_j)S_j = \hat{D}_j + \hat{V}_j$$

where

$$\hat{D}_j = -i\underline{w}\frac{\partial}{\partial\underline{\theta}} + \hat{\beta}_j \begin{pmatrix} 1 & 0 \\ 0 & -1 \end{pmatrix}$$

$$\hat{\beta}_j = \frac{1}{2}\left(\underline{wn}_j^* + 2\beta_j\right)$$

From the resonance condition we had $|\hat{\beta}_j| < \varepsilon_j$, which combined with 2.5 implies

$$\left|\underline{nw} - 2\hat{\beta}_j\right| > \varepsilon_j$$

since the $\underline{nw}$ term dominates. The resonance has thus disappeared, and one can now apply as the second step the usual method and calculate a transformation $\hat{R}_j$ by an equation of the form 2.3. We write the combination as

$$R_j = \hat{R}_j S_j$$

The iteration step can thus always be performed, with or without resonance. The last step is to put together all the transformations

$$R = \ldots R_j \ldots R_2 R_1$$

The only problem left is to determine whether this composition converges to a well-defined transformation R. The question of this convergence is nontrivial because the transformation S_j needed in case of resonance is not close to the identity. There will certainly be convergence if there are only *finitely many* R_j's with resonance. However if there are infinitely many steps with resonance, a lack of convergence is expected. In fact Eliasson has proven for a similar problem that when there are infinitely many steps with resonance the quasienergy spectrum is generically continuous [12]. Therefore it is necessary to determine for which values of the parameters $\underline{w}$ and β_0 the iteration has only finitely many resonant steps. This information can be obtained by analyzing the form of the coefficient β_k appearing in the diagonal operator obtained after k steps:

$$D_k = -i\underline{w}\frac{\partial}{\partial\underline{\theta}} + \beta_k \begin{pmatrix} 1 & 0 \\ 0 & -1 \end{pmatrix}$$

We define a new quantity

$$\rho_k = \beta_k - \frac{1}{2}\sum_{j=1}^{k} \underline{w}\hat{n}_j$$

where

$$\hat{\underline{n}}_j = \begin{cases} \underline{n}_j^* & \text{if there is resonance at the j-th step} \\ 0 & \text{if there isn't resonance} \end{cases}$$

The quantity ρ_k can be interpreted as the coefficient of the diagonal but subtracting the contributions generated from the transformations S_j that eliminate the resonances. The limit $\rho = \lim_{k\to\infty}\rho_k$ is always well-defined. One can easily show that if ρ is either a rational or a diophantine number then the sum $\sum_{j=1}^{k}\hat{n}_j$ is necessarily a finite integer, and as a consequence there are only finitely many resonant steps. Thus for almost all values of ρ (i.e. except for a subset of measure zero) there are only finitely many resonances and the composition 2.1 converges. What is left to show is that this set of almost all ρ's translates into a set of almost all values of the parameters $\underline{w}$ and β_0. The case of fixed $\underline{w}$ is treated in [13]. For fixed β this property can be deduced [14] from an generalization of Cassel's lemma [15], and one obtains the following result:

• For fixed β and almost all $\alpha = \omega_1/\omega_2$ there is an $\varepsilon_0(\alpha)$ such that $\forall\varepsilon < \varepsilon_0$ the quasienergy spectrum is pure point.

Acknowledgements: I would like to thank R. Krikorian for helpful discussions.

References

1. G. Casati, L. Molinari; Progr. Theo. Phys. Suppl. **98** (1989) 287.
2. J. Bellissard; in *Trends in the Eighties*, edited by S. Albeverio and Ph. Blanchard, World Scientific, Singapore 1985.
3. H.R. Jauslin; Stability and chaos in classical and quantum Hamiltonian systems; in P.L. Garrido, J. Marro (eds.), *II Granada Lectures in Computational Physics*, World Scientific, Singapore 1993.
4. M. Samuelides, R. Fleckinger, L. Touzillier, J. Bellissard; Europhys. Lett. **1** (1986) 203.
5. J.M. Luck, H. Orland, U. Smilansky; J. Stat.Phys. **53** (1988) 551.
6. P. Blekher, H.R. Jauslin, J.L. Lebowitz; Floquet spectrum for two-level systems in quasiperiodic time dependent fields, J. Stat.Phys. **68** (1992) 271.
7. H.R. Jauslin, J.L. Lebowitz; Chaos 1 (1991) 114.
8. T.S. Ho, S.I. Chu; J.Phys. **B17** (1984) 2101.

9.   T. Kato; *Perturbation Theory for Linear Operators*, Springer, Berlin 1980.
10. M. Combescure; Ann. Inst. H. Poincaré **44** (1986) 293, and Ann. Phys. **173** (1987) 210.
11. L.H. Eliasson; Floquet solutions for the 1-dimensional quasi-periodic Schrödinger equation, Commun. Math. Phys.**146** (1992) 447.
12. L.H. Eliasson; Ergodic Skew-systems on $\mathbf{T}^d \times SO(3, \mathbf{R})$, preprint 1991.
13. R. Krikorian; in preparation.
14. H.R. Jauslin; in preparation.
15. J.W.S. Cassels; *An Introduction to Diophantine Approximation*, Cambridge University Press, 1957; and Proc. Cambridge Phil. Soc. **46** (1950) 209.

DISSIPATION AND NOISE IN QUANTUM MECHANICS

N.G. VAN KAMPEN
Institute for Theoretical Physics,
University of Utrecht, Netherlands

Abstract. First a quantum Langevin equation is constructed by adding a noise term to the Schrödinger equation. This leads to the general form of a master equation for the evolution of the density matrix. In order to obtain the specific form for a given system it is necessary to include the interaction with a bath. The Markov property, however, can only be introduced by assumption. Here a specific soluble model is described, in which this assumption can be tested. It appears that the interaction must be weak, and that the temperature must be so high that the interaction is virtually classical.

1. The Schrödinger–Langevin equation

Newton's equation of motion for a particle (in one dimension, mass 1) subject to a force with potential $\Phi(x)$ is

$$\ddot{x} = -\Phi'(x) . \tag{1}$$

Damping can be taken into account by adding a friction force:

$$\ddot{x} = -\Phi'(x) - \gamma\dot{x} , \tag{2}$$

involving a phenomenological friction coefficient γ. Subsequently Langevin taught us how to include noise by adding a fluctuation term

$$\ddot{x} = -\Phi'(x) - \gamma\dot{x} + C\,\ell(t) , \tag{3}$$

where $\ell(t)$ is a random function independent of x, and endowed with the properties

$$\langle\ell(t)\rangle = 0, \qquad \langle\ell(t)\,\ell(t')\rangle = \delta(t - t') . \tag{4}$$

(When $\Phi'(x)$ is nonlinear one needs a further assumption, e.g., that $\ell(t)$ is Gaussian.)

The noise introduces a second coefficient C, which, however, is related to γ by the fluctuation–dissipation theorem

$$\frac{1}{2}C^2 = \gamma \cdot k_B\,T . \tag{5}$$

It expresses the fact that in thermal equilibrium the energy dissipated by the friction is supplied by the noise source. Hence the noise in a system, such as an electric circuit, can be described without additional knowledge. However, the question remains whether this formal procedure is correct. This question can be decided only

131

Z. Haba et al. (eds.), Stochasticity and Quantum Chaos, 131–136.

by examining the mechanism that causes the damping and the noise, but that is not our present task [1].

Quantum mechanics replaces (1) with

$$\dot\psi(x,t) = -iH\,\psi(x,t), \qquad H = -\frac{1}{2}\,\nabla^2 + \Phi(x)\;. \tag{6}$$

One might think that the analog of (2) is

$$\dot\psi = -iH\,\psi - U\,\psi\;, \tag{7}$$

where U is an operator, which may be taken hermitian (because an anti–hermitian part could be absorbed in H) and positive definite to avoid negative damping. However, the effect of such a term would be that the norm $\|\,\psi\,\|$ decreases, i.e., probability is dissipated rather than energy, which makes no sense. On the other hand, it is possible to increase $\|\,\psi\,\|$ by means of a noise term:

$$\dot\psi = -iH\,\psi - U\,\psi + \ell(t)\,V\,\psi \tag{8}$$

with an arbitrary operator V. The two effects balance after averaging over $\ell(t)$ when

$$\frac{1}{2}V^\dagger\,V = U\;. \tag{9}$$

Hence (8) combined with (9) constitutes an admissible quantum Langevin equation.

Note that most authors who try to construct quantum analogs of the classical Langevin equation think in terms of the Heisenberg representation [2]. I shall therefore refer to (8) with (9) as the "Schrödinger–Langevin equation".

Note also that, although (9) has a similar role as the classical fluctuation-dissipation theorem (5), it is fundamentally different as it does not involve T. It is therefore not possible to suppress the noise by going to low T. Hence in quantum mechanics no analog of (2) exists.

2. The quantum master equation

A more general Schrödinger–Langevin equation obtains by adding a number of damping terms with corresponding noise terms

$$\dot\psi = -iH\,\psi - \frac{1}{2}\sum_\alpha V_\alpha^\dagger\,V_\alpha\,\psi + \sum_\alpha \ell_\alpha(t)\,V_\alpha\,\psi\;, \tag{10}$$

where V_α are arbitrary operators and

$$\langle \ell_\alpha(t)\,\ell_\beta(t')\rangle = \delta_{\alpha\beta}\,\delta(t-t')\;. \tag{11}$$

Like in the classical case the Langevin equation can be translated into a master equation. The only difference is that we do not need the entire probability distribution $P(\psi)$ in the space of functions ψ, because all information needed is present in the second moments, i.e., the density matrix

$$\rho(x, x') = \langle \psi(x)\,\psi^*(x')\rangle\;. \tag{12}$$

Here the angular brackets denote averages over the noise sources $\ell_\alpha(t)$. The resulting master equation is

$$\dot{\rho} = -i[H, \rho] + \sum_\alpha \left\{ V_\alpha \, \rho \, V_\alpha^\dagger - \frac{1}{2} V_\alpha^\dagger \, V_\alpha \, \rho - \frac{1}{2} \rho \, V_\alpha^\dagger \, V_\alpha \right\} \ . \tag{13}$$

This equation is well-known from another point of view. It has been shown mathematically by Kossakowski and others [3] to be the most general form of differential equation for ρ, subject to the physically obvious requirements

$$Tr \, \rho = 1 \ ; \tag{14a}$$

$$\rho^\dagger = \rho \ ; \tag{14b}$$

$$\rho \quad \text{is positive (semi--)definite} \ . \tag{14c}$$

The crucial question that remains unanswered is whether ρ does obey a differential equation of the form

$$\dot{\rho}(t) = \mathcal{L} \, \rho(t) \ . \tag{15}$$

Here $\mathcal{L}$ is a time–independent superoperator, i.e., a linear mapping of any operator ρ into another one. To put this question differently: Is the process $\rho(t)$ Markovian? Or still differently: Does $\rho(t)$ constitute a semigroup with infinitesimal generator $\mathcal{L}$? In addition there is the task of determining the specific form of the operators V_α for any given system.

3. The standard method

In order to describe for a given system S the damping and the noise one has to include the effect of the interaction with a bath B. The total Hamiltonian is

$$H_T = H_S + H_B + H_I \ . \tag{16}$$

The total density matrix evolves according to

$$\rho_T(t) = e^{-itH_T} \, \rho_T(0) \, e^{itH_T} \tag{17}$$

and the density matrix of S alone is

$$\rho_S(t) = \mathop{Tr}_B \rho_T(t) \ . \tag{18}$$

These are mathematical tautologies, but the problem is how to choose $\rho_T(0)$. Customarily one takes a direct product [4]

$$\rho_T(0) = \rho_S(0) \otimes \rho_B^e \ , \tag{19}$$

where $\rho_S(0)$ may be any initial state of S, while ρ_B^e is the equilibrium distribution of B at some bath temperature T. Then

$$\rho_S(t) = \mathop{Tr}_B \, e^{-itH_T} \, \rho_S(0) \, \rho_B^e \, e^{itH_T} \ . \tag{20}$$

This defines a map $\rho_S(0) \to \rho_S(t)$, but this map will not be a semigroup. For, if $t_1 > 0$, the map $\rho_S(t_1) \to \rho_S(t_1 + t)$ will not be the same one, since at t_1 the bath is no longer in the equilibrium ρ_B^e, owing to the preceding interaction with S. Moreover there will be correlations between B and S so that $\rho_T(t_1)$ does not factorize as in (19).

This difficulty is usually overcome by wishful thinking: one assumes that the correlations created by the past interaction do not influence the subsequent evolution of S. Accordingly one assumes (often tacitly) that at t_1 one may start again with $\rho_T(t_1) = \rho_S(t_1)\rho_B^e$. This crucial assumption permeates the whole of statistical mechanics of irreversible processes under various names: Stosszahlansatz, molecular chaos, random phase, or repeated randomness assumption. Without the aid of this assumption, applied after every short interval Δt, it would be impossible to utilize perturbation theory for computing transport coefficients. A number of authors have thought that they could eliminate it, but nowadays it is rarely mentioned in polite company. The purpose of this work is to test this assumption on a soluble model of an irreversible process.

4. The model

Free particle in one dimension with mass M; a bath consisting of a set of oscillators with frequencies k_n; and a bilinear H_I with coupling constants v_n:

$$H_T = \frac{P^2}{2M} + \sum_n k_n a_n^\dagger a_n + P \sum_n v_n \left(a_n + a_n^\dagger \right) . \tag{21}$$

Thermal equilibrium of B is given by

$$\rho_B^e = \prod_n \left(1 - e^{-\beta k_n} \right) e^{-\beta k_n a_n^\dagger a_n} . \tag{22}$$

Since H_T commutes with P it is diagonal in the P-representation; its matrix elements are

$$
\begin{aligned}
(P |H_T| P') &= \delta (P - P') \left[\frac{P^2}{2M} + \sum_n \left\{ k_n a_n^\dagger a_n + P v_n \left(a_n + a_n^\dagger \right) \right\} \right] \\
&= \delta (P - P') \left[\frac{P^2}{2M} + \sum_n H_n(P) \right] .
\end{aligned}
\tag{23}
$$

The $H_n(P)$ operate in the space of B and depend parametrically on P.

In this representation the equation (20) reduces to

$$
\begin{aligned}
(P |\rho_S(t)| P') &= (P |\rho_S(0)| P') \exp \left[\frac{P^2 - P'^2}{2iM} t \right] \\
&\times \prod_n \left[\left(1 - e^{-\beta k_n} \right) \, Tr \left\{ e^{-itH_n(P)} \, e^{-\beta k_n a_n^\dagger a_n} \, e^{itH_n(P')} \right\} \right] .
\end{aligned}
\tag{24}
$$

This expression can be worked out explicitly [5] to give

$$(P \,|\rho_S(t)|\, P') = (P \,|\rho_S(0)|\, P') \, \exp \left[\frac{P^2 - P'^2}{2iM'} \, t \right]$$

$$\times \exp \left[iF(t) \left(P^2 - P'^2 \right) \right] \, \exp \left[-G(t) \left(P - P' \right)^2 \right] . \tag{25}$$

Here M' is a renormalized mass, and F and G are functions of t, well–defined in terms of the coefficients k_n, v_n but a bit complicated.

It can be verified explicitly that (25) obeys the requirements (14), but of course this is a priori obvious as (25) is exact.

5. The Markov character

In order that (25) be a semigroup it has to be true that for $t_1, t_2 > 0$

$$F(t_1 + t_2) = F(t_1) + F(t_2) , \qquad G(t_1 + t_2) = G(t_1) + G(t_2) . \tag{26}$$

Thus F and G ought to be proportional to t – which they are not. However, it is shown in [5] that they are approximately proportional in the following sense.

There exists a transient time ϑ, determined by the properties of the bath, such that for $t > \vartheta$ one has approximately

$$F(t) = \Gamma \, t, \qquad G(t) = \frac{\pi \gamma}{\beta} \, t . \tag{27}$$

with constants Γ, γ, which are expressions in the k_n, v_n. By choosing the k_n, v_n suitably one can reduce ϑ except for the following restriction:

$$\vartheta^{-1} \ll k_B T \Big/ h. \tag{28}$$

This restriction is independent of the particular bath and involves only its temperature. The right–hand side is the frequency at which quantum effects become noticeable. The conclusion is that there is no Markov character on a time scale on which the quantum character of the bath oscillators is relevant. Quantum noise is not white. The quantum correction to the classical Nyquist formula [6] has no physical meaning.

Accepting the existence of ϑ, one may subdivide the t–axis into successive intervals Δt, large compared to ϑ. For each Δt one may apply (25) with F and G given by (27). These intervals can be chained together to form a discrete-time Markov process. This discretized evolution will be close to the real one provided that the error made during each Δt is small. That amounts to the requirement that the interaction must be sufficiently weak to ensure that the change in $\rho_S(t)$ during each Δt is small.

References

1. N.G. van Kampen, *Stochastic Processes in Physics and Chemistry* (North–Holland, Amsterdam 1981, 1992).

2. E.g. C.W. Gardiner, *Quantum Noise* (Springer, Berlin 1991)
3. A. Kossakowski, Bull. Acad. Pol. Sci., Ser. Math. Astr. Phys. **20**, 1021 (1971); **21**, 649 (1973);
    V. Gorini, A. Kossakowski, and E.C.G. Sudarshan, J. Math. Phys. **17**, 821 (1976);
    G. Lindblad, Commun. Math. Phys. **40**, 147 (1975)
4. W.H. Louisell, *Quantum Statistical Properties of Radiation* (Wiley, New York 1973) ch. 6.
5. N.G. van Kampen, to be published
6. H. Nyquist, Phys. Rev. **32**, 110 (1928)

IRREGULAR SCATTERING, NUMBER THEORY, AND STATISTICAL MECHANICS

ANDREAS KNAUF

Technische Universität, Fachbereich 3 - Mathematik, MA 7-2, Strasse des 17.
Juni 135, D-10623 Berlin, Germany. e-mail: knauf@math.tu-berlin.de

1. Introduction

In this paper we present results on the relations between wave scattering in the so-called modular domain, analytic number theory, and classical statistical mechanics. The motion in the modular domain is an example of so-called arithmetic chaos. In this paper we shortly review results from [5, 6, 7], but concentrate on previously unpublished results.

The *modular group* $\Gamma \equiv \mathrm{PSL}(2, \mathbb{Z}) := \mathrm{SL}(2, \mathbb{Z})/\{\pm 1\}$ acts on the *upper half plane* $\mathbb{H} := \{z \in \mathbb{C} \mid \mathrm{Im}(z) > 0\}$ by *Möbius transformations*:

$$\hat{M}(z) := \frac{az + b}{cz + d},$$

$M = \begin{pmatrix} a & b \\ c & d \end{pmatrix} \in \mathrm{SL}(2, \mathbb{Z})$ being a determinant one matrix with integer entries.

The *modular domain* (see Fig. 1)

$$\mathbb{F} := \{z \in \mathbb{H} \mid -\tfrac{1}{2} < \mathrm{Re}(z) < \tfrac{1}{2}, |z| > 1\}$$

is a *fundamental domain* of that group action, that is, every $z \in \mathbb{H}$ can be carried by $\hat{M} \in \Gamma$ to a point $\hat{M}(z) \in \bar{\mathbb{F}}$, and no point of $\mathbb{F}$ can be carried to another point of $\mathbb{F}$ in such a way. The upper half plane is endowed with the Riemannian metric

$$ds^2 = \frac{dx^2 + dy^2}{y^2}, \quad z \equiv x + iy \in \mathbb{H} \tag{1}$$

of curvature -1, which is invariant under arbitrary Möbius transformations.

As can be seen from (1), the Laplace-Beltrami operator Δ on $\mathbb{H}$ has the form $\Delta = y^2(\partial_x^2 + \partial_y^2)$. We shall analyze the non-Euclidean wave equation

$$\frac{\partial^2 u}{\partial t^2} = \left(\Delta + \frac{1}{4}\right) u \tag{2}$$

restricted to the modular domain $\mathbb{F}$. In other words, we consider initial data for (2) which are *automorphic functions* $f : \mathbb{H} \to \mathbb{C}$, $f \circ \hat{M} = f$ for $\hat{M} \in \Gamma$. Then by Γ-invariance of Δ the time-t solutions are automorphic, too.

137

Z. Haba et al. (eds.), Stochasticity and Quantum Chaos, 137–148.
© 1995 *Kluwer Academic Publishers.*

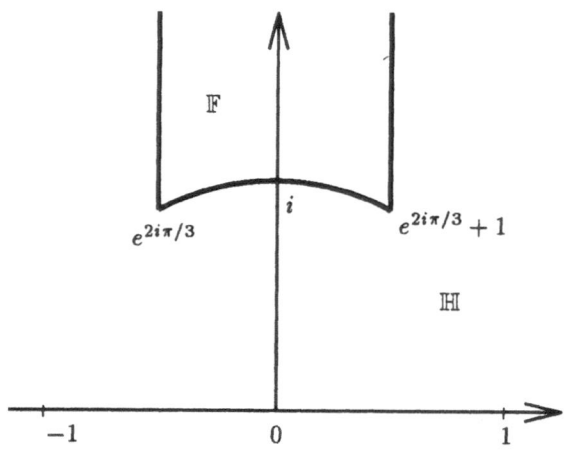

Fig. 1. The modular domain $\mathbb{F}$

In [3] Fadeev and Pavlov worked out the scattering theory on $\mathbb{F}$, that is, the automorphic solutions of (2) which come from and return to the cusp at $z = i\infty$. They showed that the scattering matrix S has the form

$$S(\omega) = \frac{\Gamma(\frac{1}{2})\Gamma(i\omega)}{\Gamma(\frac{1}{2} + i\omega)} \cdot \frac{\zeta(2i\omega)}{\zeta(1 + 2i\omega)}, \tag{3}$$

ζ being the Riemann zeta function.

In [8, 9] Lax and Phillips analyzed scattering in the modular domain and related questions in the context of their abstract scattering theory.

The underlying classical system being geodesic motion, it should not come as a surprise that (3) has a semiclassical interpretation in terms of geodesics coming from and going to the cusp. This interpretation is presented in Sect. 2. What may be more astonishing is that semiclassics can be described in terms of a classical spin chain. In Sect. 3 we describe the interaction between the spins in that chain and present some results on its thermodynamic behaviour.

In the last section, we discuss the question whether that thermodynamic description can lead to new results in scattering theory and in number theory.

2. Scattering in the Modular Domain

In this section we give a short review of the derivation of the scattering matrix (3) and interpret it semiclassically.

The x-independent solutions of the wave equation (2) are of the form

$$u(z, t) = \sqrt{y} \cdot \left(u_-(\ln(y) + t) + u_+(\ln(y) - t) \right).$$

In particular, one has plane waves of frequency ω

$$u_\omega(z,t) = \sqrt{y} \cdot \left(c_- e^{i\omega(\ln(y)+t)} + c_+ e^{i\omega(-\ln(y)+t)} \right). \tag{4}$$

As remarked in Balazs and Voros [2], $\ln(y) = \int_{z=i}^{z=iy} ds$ is the classical action and $\sqrt{y}$ the square root of the invariant density, so that (4) is a semiclassical (WKB) wave.

We now construct automorphic generalized solutions of the wave equation by starting from $h_\omega(z) := y^{\frac{1}{2}+i\omega}$. h_ω is already invariant under the subgroup $\Gamma_\infty \subset \Gamma$ of integer translations $z \mapsto z + n$. So we only sum over the right cosets $\Gamma_\infty \setminus \Gamma$ and obtain the *Eisenstein series*

$$e(z,\omega) := \sum_{\hat{M} \in \Gamma_\infty \setminus \Gamma} h_\omega(\hat{M}(z)) \tag{5}$$

which is known to converge for frequencies ω with $\text{Im}(\omega) < -\frac{1}{2}$. By construction (5) is automorphic w.r.t. $z \in \mathbb{H}$. The scattering matrix $\mathcal{S}(\omega)$ is defined with the help of the zeroth x-Fourier mode

$$e^{(0)}(y,\omega) := \int_{-\frac{1}{2}}^{\frac{1}{2}} e(x+iy,\omega)dx, \qquad y > 0,$$

namely

$$e^{(0)}(y,\omega) = y^{\frac{1}{2}+i\omega} + \mathcal{S}(\omega)y^{\frac{1}{2}-i\omega}. \tag{6}$$

If $M = \begin{pmatrix} a & b \\ c & d \end{pmatrix} \in \text{SL}(2,\mathbb{Z})$ is not in the coset space of the identity, then $c \neq 0$ and we can assume $c > 0$ because we are dealing with $\Gamma = \text{SL}(2,\mathbb{Z})/\{\pm 1\}$. Then the other elements of the right coset of M are of the form $\begin{pmatrix} a-nc & b-nd \\ c & d \end{pmatrix}$, $n \in \mathbb{Z}$, so that there is a unique representative with $0 < a \leq c$. The greatest common divisor $(c,d) = 1$, since $ad - bc = 1$. If, on the other hand, we are given $c > 0$ and d with $(c,d) = 1$, then there is a unique $M = \begin{pmatrix} a & b \\ c & d \end{pmatrix} \in \text{SL}(2,\mathbb{Z})$ with $0 < a \leq c$. Furthermore,

$$h_\omega(\hat{M}(z)) = \left(\text{Im}\left(\frac{az+b}{cz+d} \right) \right)^{\frac{1}{2}+i\omega} = \frac{y^{\frac{1}{2}+i\omega}}{[(cx+d)^2 + c^2y^2]^{\frac{1}{2}+i\omega}},$$

so that one need not determine the integers a and b. Thus one gets

$$e(z,\omega) = y^{\frac{1}{2}+i\omega} \left(1 + \sum_{c \in \mathbb{N}} \sum_{d \in \mathbb{Z},(c,d)=1} [(cx+d)^2 + c^2y^2]^{-\frac{1}{2}-i\omega} \right)$$

and

$$e^{(0)}(y,\omega) = y^{\frac{1}{2}+i\omega} \left(1 + \sum_{c \in \mathbb{N}} \sum_{d \in \mathbb{Z},(c,d)=1} \int_{-\frac{1}{2}}^{\frac{1}{2}} [(cx+d)^2 + c^2y^2]^{-\frac{1}{2}-i\omega} dx \right)$$

$$= y^{\frac{1}{2}+i\omega} \left(1 + \sum_{c \in \mathbb{N}} \sum_{1 \leq d \leq c,(c,d)=1} \int_{-\infty}^{\infty} [(cx+d)^2 + c^2y^2]^{-\frac{1}{2}-i\omega} dx \right)$$

$$= y^{\frac{1}{2}+i\omega} + y^{\frac{1}{2}-i\omega} \frac{\Gamma(\frac{1}{2})\Gamma(i\omega)}{\Gamma(\frac{1}{2}+i\omega)} \sum_{c \in \mathbb{N}} \frac{\varphi(c)}{c^{1+2i\omega}} \tag{7}$$

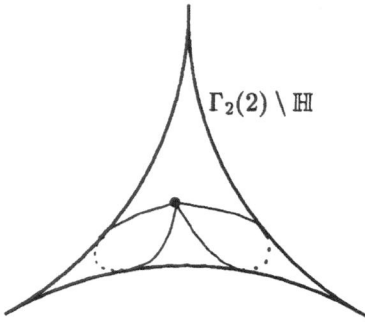

Fig. 2.   An embedding of the surface $\Gamma_2(2) \setminus \mathbb{H}$ in $\mathbb{R}^3$, with generators of its fundamental group

with the Euler totient $\varphi(c) = \#\{d \in \{1, \ldots, c\} \mid (c, d) = 1\}$.

By (6) this implies the Fadeev-Pavlov formula (3) for the scattering matrix $\mathcal{S}(\omega)$, since the quotient $\zeta(s-1)/\zeta(s)$ has the Dirichlet series

$$\frac{\zeta(s-1)}{\zeta(s)} = \sum_{n \in \mathbb{N}} \frac{\varphi(n)}{n^s}.$$

Now we will give a semiclassical geometrical interpretation of the Fadeev-Pavlov formula in terms of geodesics.

When we identify corresponding points on the boundary of $\bar{\mathbb{F}}$, we obtain the Riemannian surface $\Gamma \setminus \mathbb{H}$ which is of finite volume. But the action of the modular group Γ on $\mathbb{H}$ is not free: the point $i \in \mathbb{H}$ is a fixed point of the involutive transformation $z \mapsto -1/z$, and similarly $\hat{S}(e^{2\pi i/3} + 1) = e^{2\pi i/3} + 1$ for $S = \begin{pmatrix} 0 & 1 \\ -1 & -1 \end{pmatrix} \in \mathrm{SL}(2, \mathbb{Z})$, S generating an order three subgroup. If we try to endow $\Gamma \setminus \mathbb{H}$ with the metric coming from $\mathbb{H}$, this leads to conical singularities.

If we want to get rid of these problems, we may consider a smooth six-fold covering surface of $\Gamma \setminus \mathbb{H}$ which has the form $\Gamma_2(2) \setminus \mathbb{H}$, where $\Gamma_2(n) \subset \Gamma$ denotes the subgroup whose elements have matrix representations M of the form $M = \begin{pmatrix} 1 & 0 \\ 0 & 1 \end{pmatrix} \pmod{n}$. $\Gamma_2(2)$ is of index 6 in Γ and is freely generated by two generators with the matrices $\begin{pmatrix} 1 & 2 \\ 0 & 1 \end{pmatrix}$ and $\begin{pmatrix} 1 & 0 \\ 2 & 1 \end{pmatrix}$.

$\Gamma_2(2)$ acts freely on $\mathbb{H}$ so that $\Gamma_2(2) \setminus \mathbb{H}$ is a smooth Riemannian manifold of area 2π and curvature -1. It has three cusps (say, $i\infty$, $\hat{S}(i\infty)$ and $\hat{S}^{-1}(i\infty)$), and it is homeomorphic to a sphere with three points deleted.

In Figure 2 we show $\Gamma_2(2) \setminus \mathbb{H}$ together with representatives of the two generators of its fundamental group. Although the embedding of the surface in $\mathbb{R}^3$ shown in Fig. 2 is not isometric (probably such an embedding does not exist), it is relatively easy to visualize geodesic motion on it.

There is another interesting aspect concerning $\Gamma_2(2) \setminus \mathbb{H}$. Namely, its fundamental group, a free group on two generators, is the same as that of a certain Riemann surface that was used in [4] in order to describe the motion of a particle in the field of three nuclei. It is thus an open question whether parts of the analysis given in this

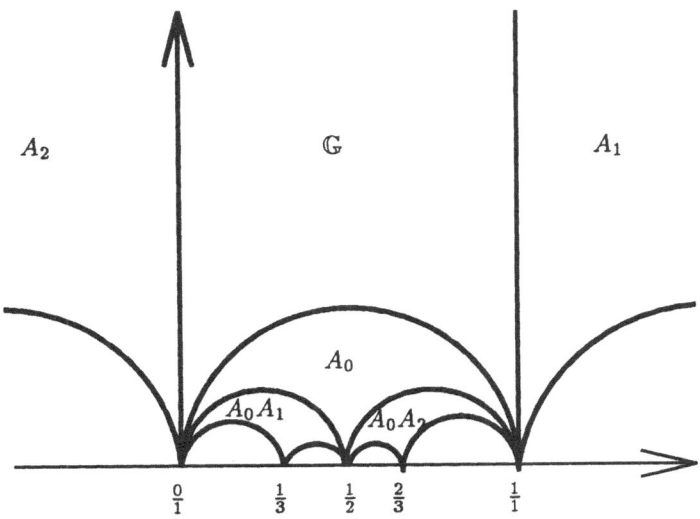

Fig. 3.   The fundamental domain $\mathbb{G}$ of the Farey tesselation

paper carry over to the more physical situation of planar scattering by Coulombic potentials.

However, it turns out that for our purposes it is more useful to consider still another tesselation of the upper half plane, starting with the fundamental domain

$$\mathbb{G} := \{z \in \mathbb{H} \mid 0 < \mathrm{Re}(z) < 1, |z - \tfrac{1}{2}| > \tfrac{1}{2}\},$$

see Fig. 3.

$\mathbb{G}$ has three times the volume of $\mathbb{F}$ and the Möbius transformation $\hat{S}$ leads to a cyclic permutation of the three cusps at $i\infty$, 0 and 1.

$\mathbb{G}$ is the fundamental domain of the subgroup of Γ which is generated by the order two transformations $\hat{A}_0, \hat{A}_1, \hat{A}_2 \in \Gamma$ with matrices

$$A_0 := \begin{pmatrix} 1 & -1 \\ 2 & -1 \end{pmatrix}, \quad A_1 := S A_0 S^{-1} = \begin{pmatrix} 1 & -2 \\ 1 & -1 \end{pmatrix} \text{ and } A_2 := S^2 A_0 S^{-2} = \begin{pmatrix} 0 & -1 \\ 1 & 0 \end{pmatrix}.$$

Because of its connection with Farey fractions C. Series christened the corresponding tesselation the Farey tesselation [10].

Proposition 1 *Every $M \in \mathrm{SL}(2, \mathbb{Z})$ can be uniquely written in the form*

$$M = \pm A_{\pi(1)} \cdot \ldots \cdot A_{\pi(k)} \cdot S^j \tag{8}$$

with $\pi(n) \in \{0, 1, 2\}$, $\pi(n) \neq \pi(n+1)$, and $j \in \{0, 1, 2\}$ (For $k = 0$ this notation means $M = \pm S^j$).

In the Appendix we give an algorithm for obtaining the above decomposition and, using that algorithm, uniqueness follows.

Corollary 2 *For $M \in \mathrm{SL}(2,\mathbb{Z})$ the domain $\hat{M}(\mathbb{G})$ is contained in*
0. $\{z \in \mathbb{H} \mid 0 < \mathrm{Re}(z) < 1\}$ iff $M = \pm S^j$ or $\pi(1) = 0$
1. $\{z \in \mathbb{H} \mid 1 < \mathrm{Re}(z)\}$ iff $\pi(1) = 1$
2. $\{z \in \mathbb{H} \mid \mathrm{Re}(z) < 0\}$ iff $\pi(1) = 2$
in the decomposition (8).

Proof. Since $\mathbb{G}$ is bounded by three geodesics ending in the cusps at ∞, 0 and 1, it suffices to observe that $\hat{A}_m(I_l) \subset I_m$, $m \neq l$, for the intervals $I_0 :=\,]0,1[$, $I_1 :=\,]1,\infty[$ and $I_2 :=\,]-\infty, 0[$ between these points.

Then the proof follows by induction in $k \in \mathbb{N}$ from Proposition 1. $\square$

From Prop. 1 and Cor. 2 it follows that there is a one-to-one correspondence between translates $\hat{M}(\mathbb{G}) \subset \{z \in \mathbb{H} \mid 0 < \mathrm{Re}(z) < 1\}$ and matrices in

$$\mathcal{U} := \left\{ \begin{pmatrix} a & b \\ c & d \end{pmatrix} \in \mathrm{SL}(2,\mathbb{Z}) \mid 0 \leq a \leq c, 0 \leq b \leq d \right\} \cup \{\mathbb{1}\}.$$

On the other hand, denoting by

$$\mathcal{U}_{\mathrm{LP}} := \left\{ \begin{pmatrix} a & b \\ c & d \end{pmatrix} \in \mathrm{SL}(2,\mathbb{Z}) \mid 1 \leq a \leq c, 1 \leq d \leq c \right\}$$

the class of matrices over which the summation in (7) is performed, we have the isomorphism

$$\mathcal{U}_{\mathrm{LP}} \to \mathcal{U}, \qquad \begin{pmatrix} a & b \\ c & d \end{pmatrix} \mapsto \begin{pmatrix} a-b & b \\ c-d & d \end{pmatrix}.$$

Therefore, we may rewrite the sum appearing in (7)

$$\sum_{c \in \mathbb{N}} \sum_{1 \leq d \leq c,\, (c,d)=1} c^{-1-2i\omega} = \sum_{\begin{pmatrix} a & b \\ c & d \end{pmatrix} \in \mathcal{U}} (c+d)^{-1-2i\omega}$$

$$= \sum_{\begin{pmatrix} a & b \\ c & d \end{pmatrix} \in \mathcal{U}} \exp(-2\ln(c+d) \cdot (\tfrac{1}{2} + i\omega)). \qquad (9)$$

This substitution has a simple semiclassical interpretation.

To every $M = \begin{pmatrix} a & b \\ c & d \end{pmatrix} \in \mathcal{U}$ we associate the vertical geodesic with real part $(a+b)/(c+d)$ coming from the middle of the three cusps of $\hat{M}(\mathbb{G})$, see Fig. 4a).

The Ford circle (see, e.g., [1]) in $\mathbb{H}$ with center at $\frac{a+b}{c+d} + \frac{i}{2(c+d)^2}$ and radius $\frac{1}{2}(c+d)^{-2}$ is the image of the horizontal line $\mathrm{Im}(z) = 1$ under the Möbius transformation $z \mapsto \hat{M}\hat{S}^2(z)$, $M = \begin{pmatrix} a & b \\ c & d \end{pmatrix}$. Equivalently, it is the image of the Ford circle $|z - (1 + i/2)| = \frac{1}{2}$ under the transformation $z \mapsto \hat{M}(z)$. Thus it is a curve of constant phase w.r.t. the partial wave originating in $\frac{a+b}{c+d}$. Its vertical metric distance to that line equals $2\ln(c+d)$, see Fig. 4b).

This means that (9) is a sum over partial waves associated to the geodesics going to the cusp. The factors $\exp(-2\ln(c+d) \cdot (\frac{1}{2} + i\omega))$ are coming from the time delays of these partial waves relative to the partial wave emanating at $z = 1$.

Moreover, Prop. 1 suggests a partition of these partial waves into families of 2^k members. This leads us to classical statistical mechanics.

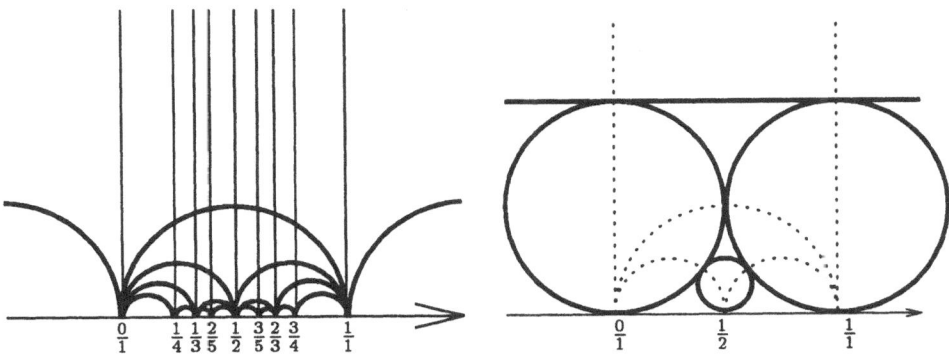

Fig. 4.   a) geodesics related to the Farey tesselation. b) The Ford circles

3. The Number-Theoretical Spin Chain

In [5] we analysed the spin chain with k spins and configuration space $\mathbf{G}_k := \mathbb{Z}_2^k$, $\mathbb{Z}_2 = \{0, 1\}$, whose canonical energy function $\mathbf{H}_k^C : \mathbf{G}_k \to \mathbb{R}$ is uniquely defined as follows: $\mathbf{H}_k^C := \ln(\mathbf{h}_k^C)$ with $\mathbf{h}_0^C := 1$, $\mathbf{h}_{k+1}^C(\sigma_1, \ldots, \sigma_k, 0) := \mathbf{h}_k^C(\sigma_1, \ldots, \sigma_k)$ and $\mathbf{h}_{k+1}^C(\sigma_1, \ldots, \sigma_k, 1) := \mathbf{h}_k^C(\sigma_1, \ldots, \sigma_k) + \mathbf{h}_k^C(1 - \sigma_1, \ldots, 1 - \sigma_k)$.

Considering the set $\mathcal{U}_k \subset \mathcal{U}$ containing the 2^k matrices of the form $M = \pm A_{\pi(1)} \cdot \ldots \cdot A_{\pi(n)} \cdot S^j$ with $n \leq k$ and $\pi(1) = 0$ if $n > 0$, we have

$$\sum_{\left(\begin{smallmatrix} a & b \\ c & d \end{smallmatrix}\right) \in \mathcal{U}_k} (c + d)^{-s} = \sum_{\sigma \in \mathbf{G}_k} \left(\mathbf{h}_k^C(\sigma)\right)^{-s} \equiv \sum_{\sigma \in \mathbf{G}_k} \exp\left(-s \cdot \mathbf{H}_k^C(\sigma)\right).$$

Thus (9) can be formally interpreted as the partition function of an infinite spin chain for inverse temperature $s = \frac{1}{2} + i\omega$.

In [5] and [7] it was shown that this interpretation is not only formal. The canonical interaction coefficients $j_k^C(t)$, $t = (t_1, \ldots, t_k) \in \mathbb{G}_k^* \cong \mathbf{G}_k$ are given by the negative Fourier transform of the canonical energy function $\mathbf{H}_k^C$:

$$j_k^C(t) := -2^{-k} \sum_{\sigma \in \mathbf{G}_k} \mathbf{H}_k^C(\sigma) \cdot (-1)^{\sigma \cdot t}.$$

Then $\mathbf{H}_k^C(\sigma) = -\sum_{t \in \mathbf{G}_k} j_k^C(t) \cdot (-1)^{\sigma \cdot t}$.

In [5] it was shown that the interaction is asymptotically translation invariant. The interaction is ferromagnetic, that is, $j_k^C(t) \geq 0$ for $t \neq 0$. The individual coefficients decay exponentially w.r.t. the distance of the spins involved in the interaction, but multi-body interactions play an important role.

The effective interaction decays like the squared inverse distance of the spins. It is well-known that for spin chains a faster polynomial decay of the interaction leads to a thermodynamic behaviour without phase transitions. In our case, however,

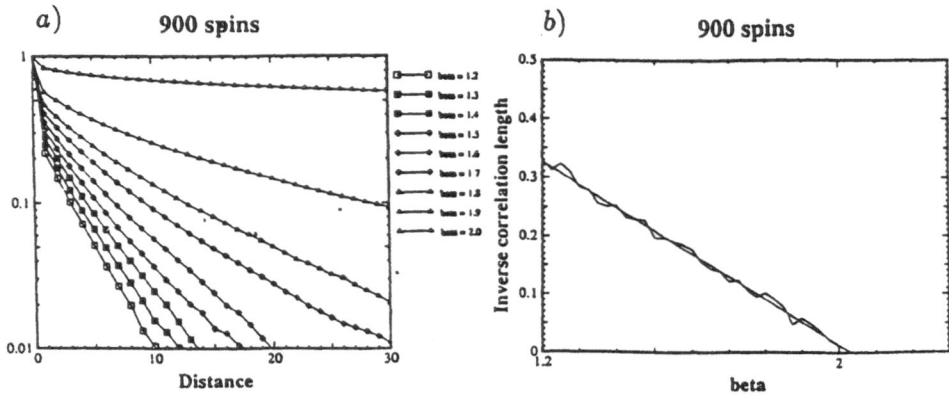

Fig. 5.   a) Two point correlation function. b) Inverse correlation length

a phase transition occurs, as shown in [6]. In [7] it was shown that for $\beta > 0$ the thermodynamic limit $F(\beta) := \lim_{k \to \infty} F_k(\beta)$ of the free energy per particle $F_k(\beta) := -\frac{1}{\beta \cdot k} \ln(Z_k(\beta))$ with partition function $Z_k(s) := \sum_{\sigma \in \mathbf{G}_k} \exp(-s \cdot \mathbf{H}_k^C(\sigma))$ exists. Since for $\mathrm{Re}(s) > 2$ the partition function of the infinite chain is finite, i.e.

$$\lim_{k \to \infty} Z_k(s) = Z(s) := \frac{\zeta(s-1)}{\zeta(s)},$$

it is clear that $F(\beta) = 0$ in the low temperature range $\beta > 2$. Moreover, the mean magnetization per spin

$$M_k(\beta) := \frac{1}{k \cdot Z_k(\beta)} \sum_{\sigma \in \mathbf{G}_k} \left(\sum_{i=1}^{k} (1 - 2\sigma_i) \right) \exp(-\beta \cdot \mathbf{H}_k^C(\sigma))$$

has a thermodynamic limit $M(\beta)$ with $M(\beta) = 1$ for $\beta > 2$. Thus for low temperatures, due to the long range of the interaction, the system is in a frozen state.

On the other hand, numerical and analytical arguments suggest a phase transition of Thouless type at $\beta_{\mathrm{cr}} = 2$, the mean magnetization jumping to $M(\beta) = 0$ for $\beta < 2$ and the energy per spin increasing sharply.

Due to the ferromagnetic character of the interaction all correlation functions are positive.

In Fig. 5a) we show a Monte Carlo calculation of the two point correlation function for a chain of $k = 900$ spins, averaged over all spins of a fixed distance. For $\beta < 2$ the correlations seem to decay exponentially w.r.t. the distance.

In Fig. 5b) the inverse correlation length is plotted as a function of β, together with a least square linear fit. The data are compatible with a phase transition at $\beta_{\mathrm{cr}} = 2$ and a critical exponent $\nu = 1$.

4. The Unreasonable Effectiveness of Statistical Mechanics

In spite of much constant effort, the behaviour of the Riemann zeta function $\zeta(s)$ in the critical strip $0 < \text{Re}(s) < 1$ is not well understood. Because of the fuctional equation $\zeta(1 - s) = 2(2\pi)^{-s}\Gamma(s)\cos(\pi s/2)\zeta(s)$ the values for arguments s with $\text{Re}(s) < \frac{1}{2}$ are determined by the values for $\text{Re}(s) > \frac{1}{2}$ and, according to the Riemann Hypothesis, all ζ function zeroes with non-vanishing imaginary part should lie on the critical line $\text{Re}(s) = \frac{1}{2}$. Both the Dirichlet series and the Euler product representations are not convergent for $\text{Re}(s) \leq 1$ so that one does not have a very direct method for calculating the values of the zeta function in the critical strip.

Thus the question arises whether the statistical mechanics approach leads to a method for calculating these values. $Z(s) = \zeta(s-1)/\zeta(s)$ has a pole at $s = 2$ which on the level of statistical mechanics probably gives rise to the above-mentioned phase transition. Thus it seems hopeless to find a convergent approximation to $Z(s)$ valid for $\text{Re}(s) > 3/2$. But this is not necessarily so.

In the theory of Dirichlet series the notion of Bohr equivalence is important. Two Dirichlet series $\sum_{n=1}^{\infty} a(n) \cdot n^{-s}$ and $\sum_{n=1}^{\infty} b(n) \cdot n^{-s}$ are Bohr equivalent iff there exists a completely multiplicative function $f : \mathbb{N} \to \mathbb{C}$ with $b(n) = f(n) \cdot a(n)$, $n \in \mathbb{N}$, and $|f(p)| = 1$ whenever $a(n) \neq 0$ and p is a prime divisor of n.

In our case $\tilde{Z}(s) := \sum_{n=1}^{\infty} \lambda(n) \cdot \varphi(n) \cdot n^{-s}$ with the Liouville function $\lambda(\prod_{p,\text{ prime}} p_i^{\alpha_i}) := (-1)^{\sum_i \alpha_i}$ is equivalent to the partition function $Z(s) = \sum_{n=1}^{\infty} \varphi(n) \cdot n^{-s}$. For $\text{Re}(s) > 2$ one has

$$\tilde{Z}(s) = \prod_{p \text{ prime}} \frac{1 + p^{-s}}{1 + p^{1-s}} = \frac{\zeta(s) \cdot \zeta(2(s-1))}{\zeta(s-1) \cdot \zeta(2s)} = \frac{Z(2s) \cdot Z(2s-1)}{Z(s)},$$

so that the analytical continuation of $\tilde{Z}(s)$ has a zero instead of a pole at $s = 2$.

Writing $Z_k(s) = \sum_{n=1}^{\infty} \varphi_k(n) \cdot n^{-s}$ with $\varphi_k(n) := \#\{\sigma \in \mathbf{G}_k \mid \mathbf{h}_k^C(\sigma) = n\}$, $\tilde{Z}(s)$ is uniformly approximated in any half-plane $\text{Re}(s) \geq 2 + \varepsilon$ by

$$\tilde{Z}_k(s) := \sum_{n=1}^{\infty} \lambda(n) \cdot \varphi_k(n) \cdot n^{-s}.$$

The question is whether one can find larger half-planes of convergence.

Because of the Riemann zeroes, $\tilde{Z}_k(s)$ has poles on the shifted critical line $\text{Re}(s) = 3/2$ so that one cannot expect convergence for s with $\text{Re}(s) \leq 3/2$.

In absence of a mathematical theory, we compared the numerical values of $\tilde{Z}_k(s)$ with $\tilde{Z}(s)$. Fig. 6a) and b) show the contour plots of $|\tilde{Z}(s)|$ and $|\tilde{Z}_{25}(s)|$ in the window $1.5 \leq \text{Re}(s) \leq 2.5$, $6 \leq \text{Im}(s) \leq 12$.

Fig. 6c) and d) show contour plots of the absolute values of the functions $\hat{Z}_m(s) := \sum_{n=1}^{m} \lambda(n) \cdot \varphi(n) \cdot n^{-s}$ for $m = 10\,000$ and $m = 250\,000$. In the first case $\hat{Z}_m$ has approximately the same number $\sum_{n=1}^{m} \varphi(n) \sim \frac{3}{\pi^2}m^2$ of terms n^{-s} as $\tilde{Z}_{25}$ (2^{25} terms). Yet, due to oscillations near the shifted critical line, the numerical accuracy of the approximation is much worse, and this accuracy does not increase much even for $m = 250\,000$. A similar behaviour is found for other values of $\text{Im}(s)$.

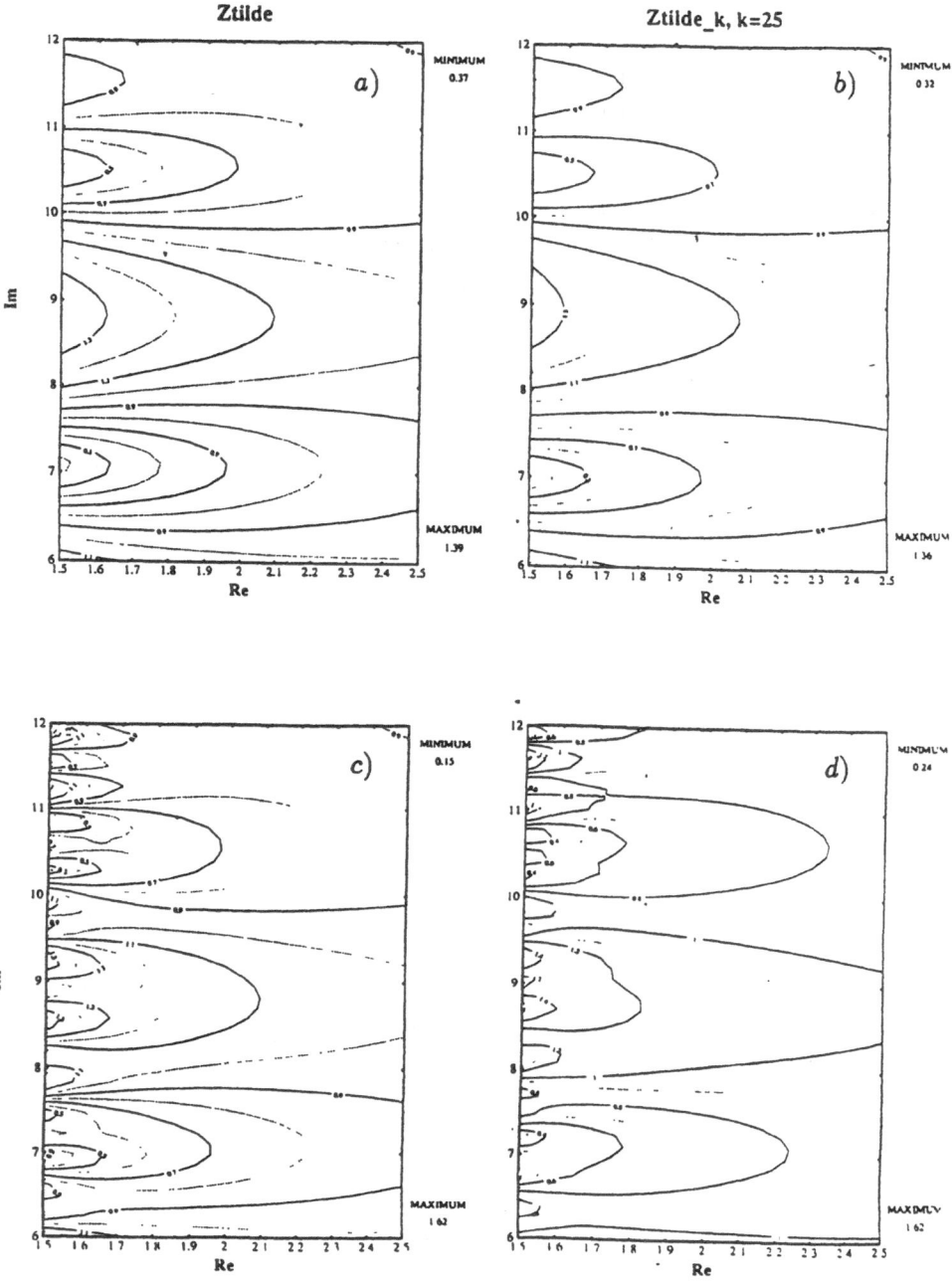

Fig. 6.   Contour plots of the absolute values of a) $\tilde{Z}(s)$ b) $\tilde{Z}_{25}(s)$ c) $\hat{Z}_{10\,000}(s)$ d) $\hat{Z}_{250\,000}(s)$

Appendix

A. An Algorithm for decomposing $\mathrm{SL}(2, \mathbb{Z})$ Elements

In this Appendix we describe an algorithm for decomposing $M \in \mathrm{SL}(2, \mathbb{Z})$ in the form $M = \pm A_{\pi(1)} \cdot \ldots \cdot A_{\pi(k)} \cdot S^j$ described in Prop. 1, using the map

$$D : \mathrm{SL}(2, \mathbb{Z}) \to \mathbb{N}, \quad \begin{pmatrix} a & b \\ c & d \end{pmatrix} \mapsto \max(|a|, |b|, |c|, |d|, |a+b|, |c+d|).$$

Obviously we have $D(MS) = D(M)$.

Lemma 3 *The subset of $M \in \mathrm{SL}(2, \mathbb{Z})$ with $D(M) = 1$ equals* $\{\pm 1, \pm S, \pm S^2, \pm A_2, \pm A_2 S, \pm A_2 S^2\}$.

Proof. By inspection. $\square$

If $D(M) = 1$, then the maximum in the definition of $D(M)$ is attained at least twice. To the contrary, for $D(M) > 1$ one has

Lemma 4 *If $D(M) > 1$, then only one element of the list $(|a|, |b|, |c|, |d| \ |a+b|, |c+d|)$ equals $D(M)$.*

Proof. We prove the statement by considering all pairs in the list:
1. The greatest common divisor of the following pairs in the list equals one: $(a, b) = (b, c) = (c, d) = (d, a) = (a, a+b) = (b, a+b) = (c, c+d) = (d, c+d) = (a+b, c+d) = 1$. If the two partners are equal, their absolute value must be one.
2. If $|a| = |d| > 1$ then $|bc| = |ad - 1| \geq |ad| - 1 = (a+1)(a-1) > 0$ which implies $\max(|b|, |c|) > |a|$ so that $|a| < D(M)$. The case $|b| = |c|$ is similar.
3. The pairs $(|a|, |c+d|)$, $(|b|, |c+d|)$, $(|c|, |a+b|)$ and $(|d|, |a+b|)$ can be transformed to the pairs of case 2) by right multiplication with $S^{\pm 1}$, remembering that $D(MS^{\pm 1}) = D(M)$. $\square$

In order to decompose M we multiply M from the *right* by the unique A_i with $D(MA_i) < D(M)$ until $D(M) = 1$. By Lemma 3 this leads to a decomposition of the form $M = \pm S^l \cdot A_{\rho(1)} \cdot \ldots \cdot A_{\rho(k)}$. Then we transform this decomposition to the form of Prop. 1, using the relation $A_i = S^i A_0 S^{-i}$.

If $D(M) > 1$ by Lemma 4 we have the three disjoint cases
0. $D(M) = \max(|a|, |c|)$
1. $D(M) = \max(|b|, |d|)$
2. $D(M) = \max(|a+b|, |c+d|)$
which are cyclically transformed into each other by the transformation $M \mapsto MS$. So we need only treat one of them, say 2). In that case we assume $D(M) = |a+b|$, the case $D(M) = |c+d|$ being similar. Then $ab > 0$ and $cd \geq 0$ (since $abcd = bc(bc+1) \geq 0$) so that $D(MA_2) = \max(|a|, |b|, |c|, |d|, |a-b|, |c-d|) < D(M)$, whereas $D(MA_0) = \max(|a + 2b|, |c + 2d|) > D(M)$ and $D(MA_1) = \max(|2a + b|, |2c + d|) > D(M)$.

So in case 2) a multiplication of M with A_2 leads to a decrease of $D(M)$. By symmetry, we multiply in the case i) with A_i. This shows uniqueness of our algorithm and also uniqueness of the decomposition (8).

References

1. Apostol, T.M.: Modular Functions and Dirichlet Series in Number Theory. Graduate Texts in Mathematics 41, New York: Springer 1990
2. Balazs, N.L., Voros, A.: Chaos on the Pseudosphere. Physics Reports **143**, 109–240 (1986)
3. Fadeev, L.D., Pavlov, B.S.: Scattering Theory and Automorphic Functions. Proc. Steklov Inst. Math. **27**, 161–193 (1972). English translation: J. Sov. Math. **3**, 522–548 (1975)
4. Klein, M., Knauf, A.: Classical Planar Scattering by Coulombic Potentials. Lecture Notes in Physics m 13. Berlin, Heidelberg, New York: Springer 1993
5. Knauf, A.: On a Ferromagnetic Spin Chain. Commun. Math. Phys. **153**, 77–115 (1993)
6. Knauf, A.: Phases of the Number-Theoretical Spin Chain. *To appear in* Journal of Statistical Physics
7. Knauf, A.: On a Ferromagnetic Spin Chain. Part II: Thermodynamic Limit. *To appear in* Journal of Mathematical Physics
8. Lax, P.D., Phillips, R.S.: Scattering Theory for Automorphic Functions. Annals of Mathematics Studies 87, New Jersey: Princeton University Press 1976
9. Lax, P.D., Phillips, R.S.: Scattering Theory for Automorphic Functions. Bulletin of the AMS (New Series) **2**, 261–295 (1980)
10. Series, C.: The Modular Surface and Continued Fractions. J. London Math. Soc. **31**, 69–80 (1985)

BAND RANDOM MATRICES, KICKED ROTATOR AND DISORDERED SYSTEMS

LUCA MOLINARI
Dipartimento di Fisica dell'Universita' di Milano and I.N.F.N.
Via Celoria 16, 20133 Milano, Italy

Abstract. An overview of the statistical properties of eigenvalues and eigenvectors of the ensemble of band random matrices (BRM) is given, showing many interesting similarities with Disordered Systems and the Kicked Rotator, a model of Quantum Chaos.

§1 Introduction;

§2 What we learn from the Kicked Rotator;

§3 About Anderson Localization;

§4 Band Random Matrices;

§5 Eigenvector statistics;

§6 Intermediate spectral statistics;

§7 The Wigner Ensemble;

§R References.

1. Introduction

The theory of Disordered Systems and Quantum Chaos are two areas of physics where static and dynamical disorder call for a statistical description, with a common reference mathematical model: random matrices. In recent years the contact has increased, and concepts like level statistics, localization, universality and scaling are becoming interdisciplinary.

The Gaussian and Circular Ensembles of random matrices [1] were introduced by Wigner and Dyson in the fifties as mathematical models mainly to describe the fluctuation properties in the observed spectra of complex physical systems, like nuclei, which resulted to be universal within some symmetry classes. Recently, the universal character of fluctuations has been recognized in systems with few degrees of freedom and chaotic classical dynamics [2][3], or in the mesoscopic fluctuations of conductance [4], or in some many-body lattice models [5].

The study of random matrices with a band structure was suggested to us by the findings of the kicked rotator (§2), a model of quantum chaos with interesting connections with disordered systems (§3). Both models are characterized by having

149

Z. Haba et al. (eds.), Stochasticity and Quantum Chaos, 149–160.
© 1995 *Kluwer Academic Publishers.*

exponentially localized eigenvectors and finite size scaling; the same properties are found for BRM (§5). The band structure is a realistic requirement for the matrix representation of Hamiltonians in a properly ordered basis, implied by the decay of correlations. From another point of view BRM appeared as the natural candidate for describing the intermediate spectral statistics (§6) of systems that are neither integrable nor completely chaotic. In the attempt to justify the random matrix approach from the generic properties of ergodic Hamiltonians, Feingold et al. (§7) introduced an ensemble of random matrices with a band structure and increasing diagonal elements, which is physically more significative to describe the transition properties. A similar ensemble of "bordered matrices" was earlier defined by Wigner [6], but then abandoned in favour of the mathematically more tractable Gaussian and Circular ensembles.

2. What we learn from the Kicked Rotator

It is instructive to review briefly some results about the "kicked rotator", a simple and rich model for quantum chaos [7] that inspired much of our work on band random matrices. The Hamiltonian is that of a free rotator, periodically kicked by an impulsive force

$$H = \frac{1}{2I}p^2 + k\cos\theta \sum_n \delta(t - nT) \tag{2.1}$$

The classical dynamics in phase space is described by the *standard map*, which gives the time evolution over one period

$$p' = p + k\sin\theta \quad , \quad \theta' = \theta + (T/I)p' \quad (\text{mod}.2\pi) \tag{2.2}$$

It is known that for $K = (kT/I) > 1$, the border for global chaos, the last invariant curve is broken, and points on average wander to higher values of p, determining an increase of the squared momentum of the rotator according to a diffusive law:

$$\langle p^2 \rangle \approx p_0^2 + Dt, \quad D = \frac{k^2}{2T} \tag{2.3}$$

In quantum mechanics, the relevant informations for the description of the dynamics of a time-periodic system are encoded in the spectral properties of the Floquet operator, the one-period time propagator. Being unitary, it has eigenvalues $\exp{-(i/\hbar)T\epsilon}$, where ϵ is a "quasienergy" value. In the angular representation the Floquet operator describing a kick followed by a free rotation is:

$$\left(\hat{F}\psi\right)(\theta) = \exp\left(i\frac{\hbar T}{2I}\frac{\partial^2}{\partial\theta^2}\right)\exp\left(-\frac{i}{\hbar}k\cos\theta\right)\psi(\theta) \tag{2.4}$$

The angular momentum is quantized, $\hat{P}|n\rangle = \hbar n|n\rangle$; in this basis, the Floquet operator has matrix elements

$$F_{m,n} = \langle m|\hat{F}|n\rangle = (-i)^{m-n}\exp\left(im^2\frac{\hbar T}{2I}\right)J_{m-n}\left(\frac{k}{\hbar}\right) \tag{2.5}$$

that decay exponentially for $|m - n|$ greater than $k/\hbar$ and, in the classical caotic regime, practically cannot be distinguished from a random sequence. In this situation, numerical computations have shown that the eigenfunctions are exponentially localized in n-space, with a localization length proportional to the square of the bandwidth of the matrix F [8][9]

$$|\psi_n|^2 \approx \exp\left(-\frac{2}{\xi_\infty}|n - n_0|\right) \quad , \quad \xi_\infty \approx \frac{1}{2}\left(\frac{k}{\hbar}\right)^2 \tag{2.6}$$

The quantum behaviour of the average squared momentum is interesting: after a transient $t < \tau$ during which it increases according to (2.3), it saturates and the wave function reaches a quasi-stationary state, exponentially localized in n-space. The length is about twice that of eigenvectors, and is approximated by $\sqrt{D\tau}/\hbar$. The break time τ can be estimated by the following argument. After a time τ, the number of n-states involved in the dynamics of an initial state $|n_0\rangle$ is $n_\tau = \sqrt{D\tau}/\hbar$. On average they have a quasi-energy spacing $\Delta\epsilon(\tau) = 2\pi\hbar(Tn_\tau)^{-1}$. Quantum effects become important as soon as $\Delta\epsilon(\tau)\tau \approx \hbar$, whence τ and ξ_∞ can be derived.

The phenomenon is known as *dynamical localization* , and has strong analogies with Anderson localization, occurring in solid state physics. It is responsible, for example, for the suppression of microwave ionization of hydrogen atoms [10]. The connection with solid state can be established through a mapping of the eigenvalue equation for the Floquet matrix on a 1D tight binding model with periodic incommensurate potential [11][8].

A different behaviour occurs in the resonant case $(\hbar T/4\pi I) = M/N$, where the Floquet matrix is periodic $F_{m,n} = F_{m+N,n+N}$. According to Bloch's construction, the eigenvalue problem is reduced to that for a $N \times N$ matrix incorporating the boundary condition $\psi_{n+N} = e^{-i\phi}\psi_n$. The full spectrum, resulting from the superposition of such spectra labelled by ϕ, is absolutely continuous and splitted in N bands. Inside a single cell of length N, and for a given Bloch phase ϕ, a localization length $\xi(N)$ of eigenvectors can be defined (§5) and is found to satisfy a scaling relation [12] that is typical of disordered systems:

$$\frac{\xi(N)}{N} = f\left(\frac{\xi_\infty}{N}\right) \tag{2.7}$$

The scaling function f is the same found for BRM (5.7). In the mapping on a tightbinding model, the resonant case corresponds to a commensurate potential.

The kicked rotator and the corresponding mapping have been generalized to D=2, where states were found localized with ξ_∞ increasing exponentially with k [13], and to D=3 where the analogous of the metallic-insulator transition was observed [13]. These behaviours are extensively investigated in a class of models for disordered transport.

3. About Anderson Localization

The low temperature tranport properties of a lattice with random impurities can be modelled through a tight binding Hamiltonian, which is the sum of a kinetic term, accounting for the electron hopping between neighboring lattice sites, and a potential, which provides disorder. The eigenvalue equation is

$$(T\psi)_{\mathbf{n}} + V_{\mathbf{n}}\psi_{\mathbf{n}} = E\psi_{\mathbf{n}} \tag{3.1}$$

where $\psi_{\mathbf{n}}$ is the amplitude at site $\mathbf{n}$, E is the energy and $\{V_{\mathbf{n}}\}$ specifes the potential. In the simplest description, the kinetic term is the sum of the amplitudes at sites nearest to $\mathbf{n}$.

In the Anderson model [15], the numbers $V_{\mathbf{n}}$ are independent random numbers with uniform density in [-W/2,W/2]; in Lloyd's model [16], to which the kicked rotator is more closely related [9], the density is Lorentzian. Another important class of disordered models is specified by potentials periodic in space, but incommensurate with the lattice period. A presentation of localization in quantum chaos and disordered systems is given by Fishman [17].

In 1D, if the hopping is short range, the eigenstates are always exponentially localized: this can be demonstrated by rewriting the recurrence equation (3.1) in a matrix form, which is iterated multiplicatively. The asymptotic properties of the eigenstates are hence determined by those of the eigenvalues of the resulting transfer matrix, for which theorems exist [18]. Localization implies a Poissonian spectral statistics; in [19] it is studied using Izrailev's distribution (6.2). Samples in 1D or 2D with infinite extension are therefore insulators, but for an infinite 3D lattice, a conducting-insulator transition occurs at a critical value of disorder. For finite size systems, the conductance is described by a scaling theory [20]. For large disorder W, when localization occurs and the decay length $\xi_{\infty}(W)$ is much smaller than the size N, conductance is exponentially small; for small disorder it obeys Ohm's law. In 1D the scaling theory for conductance is equivalent to a scaling relation for localization lengths for eigenvectors on finite samples, due to Pichard [21]

$$\frac{\xi(N, W)}{N} = f\left(\frac{\xi_{\infty}(W)}{N}\right) \tag{3.2}$$

There are various ways to study conductance in samples of finite size. In the approach by Landauer one places the sample with disorder inside a perfectly conducting chain and studies the scattering problem to relate conductance to the transmission coefficient. In the approach by Thouless, it is related to the response of energy levels to changes of boundary conditions. Such changes are produced through a phase dependence $e^{i\phi}$ of boundary conditions for eigenvectors. The second derivative of energy levels $E(\phi)$ at $\phi = 0$ defines the curvature k and has been investigated by Życzkowski et al. for the 3D Anderson model [22]. They showed that curvature, in analogy with the average conductance, obeys scaling and undergoes a transition at the same critical value of W. In the metallic regime the distribution of rescaled curvatures is universal [30]. The transition also reflects in the level spacings, as investigated by Shklovskii et al. [23].

4. Band Random Matrices

The study of the kicked rotator has led us to consider an ensemble of random matrices with a band structure. Though the Floquet operator is unitary, we considered

real symmetric and Hermitian matrices, for simplicity. Such matrices, when exponentiated, still preserve a band structure (Feingold, private comm.)
Symmetric Band Random Matrices are defined as the ensemble of real symmetric $N \times N$ matrices with nonzero matrix elements A_{ij} restricted in a band $|i - j| < b$ and distributed as independent Gaussian random variables. The number of such independent variables is $F = b(2N - b + 1)/2$. The probability densities are

$$p(A_{ii}) = \exp(-aA_{ii}^2) \quad , \quad p(A_{ij}) = \exp(-2aA_{ij}^2) \quad i < j \qquad (4.1)$$

For $b = 2$ the matrices are tridiagonal, and have been studied by Dyson [24], for $b = N$ they belong to GOE. One can also define complex or symplectic Band Random Matrices.
Although the interest in random matrices mainly lies in their fluctuation properties, providing a mathematical reference to which physical results are compared, a lot of effort was spent to investigate the distribution of eigenvalues. For the GOE ensemble, the level repulsion and the harmonic well combine to produce, for $N \to \infty$, a semicircle law ditribution for the eigenvalues, with radius R growing as $\sqrt{N/a}$

$$\rho(\lambda) = \frac{2}{\pi R^2} \sqrt{R^2 - \lambda^2} \qquad (4.2)$$

The derivation is almost trivial [1] due to the rotational symmetry that allows to factorize the measure in matrix space into an eigenvalue part, and an angular part. For Band Random Matrices this is not possible, and the analysis is very difficult. The second and fourth moments of the eigenvalue distribution are however easily calculated by computing average traces; for $b < N/2$:

$$\langle \lambda^2 \rangle = \frac{1}{4Na} b(2N - b + 1) \quad \langle \lambda^4 \rangle = \frac{1}{8Na^2} [N(4b^2 + b + 1) - \frac{5}{6} b(b - 1)(4b + 1)] \quad (4.3)$$

For large N and b, such that $b/N \to 0$, the adimensional ratio $\langle \lambda^4 \rangle / \langle \lambda^2 \rangle^2$ approaches the value 2 of the semicircle distribution. Many authors have indeed shown, by different techniques, that BRM in the above limit have a level density described by the semicircle law [25-28], with radius growing as $\sqrt{b/a}$. Note that, for large b and N, with finite ratio b/N, the limit distribution is not a semicircle.
An interesting variant of this ensemble is given by periodic $N \times N$ BRM, in which the number of matrix elements per row is kept constant, $2b - 1$, by adding an upper right and lower left triangles of non-zero random matrix elements. They may be multiplied respectively by the phase factors $e^{i\phi}$ and $e^{-i\phi}$. These matrices correspond to the Bloch decomposition of an ensemble of infinite BRM with periodicity N: $A_{i+N,j+N} = A_{i,j}$. The eigenvectors are periodic up to a phase $\psi_{j+N} = e^{-i\phi}\psi_j$ and are computed as eigenvectors of a member of periodic BRM of size N and phase ϕ. The level curvatures for periodic BRM have been studied by Casati et al. [29], finding analogies with curvatures in Anderson model. The distributions of curvatures were shown to depend only on b^2/N and agree with a universal distribution proposed by Delande and Zakrzewski [30].

5. Eigenvector Statistics

Unlike the GOE ensemble, which is invariant under rotations, band random matrices have interesting properties of eigenvectors, which can be compared to those of the Anderson model. It is useful to view a $N \times N$ BRM as the Hamiltonian for a 1D tight binding model on a lattice of length N, with random interactions of range $2b - 1$. The boundary conditions are $\psi_i = 0$ for $i \leq 0$ and $i > N$, or $\psi_{i+N} = e^{-i\phi}\psi_i$ for periodic BRM.

In the limit $N \to \infty$ the eigenvectors are exponentially localized with a localization length ξ_∞ proportional to b^2. This property, expected from the theory of the kicked rotator, was derived heuristically by Feingold et al.[31] and more rigorously by Fyodorov and Mirlin [32], who also obtained the eigenvalue dependence of localization:

$$\xi_\infty \propto b^2(R^2 - \lambda^2) \qquad |\lambda| < R \tag{5.1}$$

The last authors employ a powerful supersymmetric formalism that allows the exact evaluation of correlation functions, like the ensemble average of products of matrix elements of the resolvent.

For finite matrices, unless b is so small that the exponential decay is unaffected by the boundary, eigenstates are very irregular and become extended for $b^2 \approx N$. It is then necessary to introduce a measure of localization which looks at the bulk structure of the eigenvector, rather than the tail. An initial choice was to look at the entropy localization [33], which is a particular case of the measures defined below.

Given an ensemble of normalized eigenvectors with N components $\{u_1 \ldots u_N\}$, one defines the average moments $(q > 0)$ and the lengths, borrowed from multifractal analysis

$$P_q = \langle \sum_{n=1}^{N} |u^{2q}| \rangle \quad , \quad \xi_q = P_q^{\frac{1}{1-q}} \tag{5.2}$$

In particular, for $q = 1$ we obtain the entropy length, and for $q = 2$ the participation ratio:

$$\xi_1 = \exp\left(-\langle \sum_{i=1}^{N} u_i^2 \log u_i^2 \rangle\right) \quad ; \quad \xi_2 = \langle \sum_{i=1}^{N} |u_i^4| \rangle^{-1} \tag{5.3}$$

These quantities can be computed analytically for GOE matrices, where the invariance under rotations implies a joint probability for the eigenvector components restricted only by normalization $P(u_1 \ldots u_N) \propto \delta(1 - |u|^2)$. Integration over $N - 1$ variables gives the density for a single component:

$$P(u) = \frac{1}{\sqrt{\pi}} \frac{\Gamma(\frac{N}{2})}{\Gamma(\frac{N-1}{2})} (1 - u^2)^{\frac{1}{2}(N-3)} \tag{5.4}$$

which for $N \to \infty$ becomes the Porter Thomas distribution with the rescaling $x = u^2/N$. The moments are:

$$H^{GOE} = -\int u^2 \log u^2 P(u) du = \psi(\frac{N}{2}) - \psi(\frac{3}{2}) \to \log N - 0.7296 \tag{5.5a}$$

$$P_q^{GOE} = \int u^{2q} P(u) du = \frac{\Gamma(\frac{N}{2})\Gamma(q + \frac{1}{2})}{\Gamma(q + \frac{N}{2})\sqrt{\pi}} \rightarrow \frac{1}{\sqrt{\pi}} \left(\frac{N}{2}\right)^{1-q} \Gamma(q + \frac{1}{2}) \qquad (5.5b)$$

For large N and small q/N, the eigenstates are extended in the above measures: $\xi_q^{GOE}(N) \propto N$. Since GOE corresponds to the limit case of BRM where eigenstates are most delocalized, it is useful to introduce the ratios with values in $(0,1)$ and the normalized lengths

$$\beta_q = \frac{\xi_q^{BRM}}{\xi_q^{GOE}} \quad , \quad \ell_q = N\beta_q \qquad (5.6)$$

The entropy length ($q = 1$) was numerically found by Casati, Izrailev and Molinari [34] to satisfy a scaling relation, that can be described by a very simple scaling function. The extension to other low values of q was provided by Economou and Evangelou [35]

$$\frac{\ell_q(N, b)}{N} = f\left(c_q \frac{b^2}{N}\right) \quad , \quad f(x) = \frac{x}{1 + x} \qquad (5.7)$$

The scaling function f has the correct limit behaviour $f(x) \approx x$ for $x \rightarrow 0$ and $f(x) \rightarrow 1$ for $x \rightarrow \infty$. The scaling variable $c_q b^2/N$ should accordingly coincide with $\ell_q(\infty, b)/N$. The given form of f provides a good fitting to numerical data for b^2/N less than 10 in the case $q = 1$. Beyond this value, scaling persists but takes another form. For periodic BRM, where the effect of boundaries is removed, the deviation is absent. The relation (5.7) can be rewritten in the following interesting form:

$$\frac{1}{\xi_q(N, b)} = \frac{1}{\xi_q(N, N)} + \frac{1}{\xi_q(\infty, b)} \qquad (5.8)$$

Fyodorov and Mirlin [36] showed that the above expression of the scaling function f is approximate, and only in the case $q = 2$ it becomes exact. They were also able to derive the fluctuation properties of the inverse participation ratio P_2 [37].

These scaling properties for eigenvectors follow from a stronger scaling property: that of the whole distribution of squared components of eigenvectors in the parameter b^2/N. Zyczkowski et al. [38] showed numerically that the distribution of $x = u^2/N$, where u is any component of an eigenvector in a given energy window for BRM with sizes b and N, is the same for equal values of b^2/N. The analytical expression for the distribution $P(x, b^2/N)$ was later obtained by Fyodorov and Mirlin [39].

The scaling relation (5.7) for the entropy length was also observed in 1D Anderson and Lloyd models [40]. The scaling parameter for disordered systems, which replaces b^2/N, is $\xi_\infty(W)/N$, as appears in (3.2). The reference limit case for the normalization of lengths is the solution in the absence of disorder, since it corresponds to the most extended case. Again, the scaling of localization lengths is a consequence of the scaling of the distribution of eigenvector amplitudes, observed by Molinari [41].

6. The intermediate spectral statistics

The most investigated statistical property of spectra is the distribution of spacings between neighboring rescaled levels. The levels are rescaled in order to obtain a sequence with average unit spacing. For complex or fully chaotic systems, the distribution of spacings is well described by the Wigner-Dyson distribution

$$P(s) = As^\beta e^{-Bs^2} \tag{6.1}$$

which reproduces the exact distribution for random matrix ensembles quite well. A and B are normalization constants such that $\langle 1 \rangle = \langle s \rangle = 1$ and $\beta = 1, 2, 4$ is the parameter selected by the symmetry of the system. Integrable systems have level spacings distributed according to a Poissonian, which is the statistics for diagonal random matrices. To describe the intermediate situation, the continuous transition between the two limit statistics, Seligman, Verbaarschot and Zirnbauer [42,43] introduced an ensemble of random matrices with variable band size, as a model to interpolate between GOE and diagonal matrices. They studied the spectral statistics of two coupled anharmonic oscillators, with parameter values ranging from fully chaotic to integrable regimes. The experimental level spacing distributions and Δ_3 statistics were fitted with the corresponding level statistics obtained numerically for band random matrices of fixed and large size N, and variable bandwidth b. The agreement was found good. This approach is very empirical, the fitting value of b depending on the size N.

The analysis of level statistics of BRM was improved by Cheon [44], who showed that it is irrelevant whether the band structure is built by putting to zero all matrix elements at a distance b from the diagonal, or making them vanish exponentially on that range. He computed numerically the moments $m_k = \langle s^k \rangle$ of the level spacing distribution for various values of b and N, and found the interesting property that they all distributed along single lines in the $m_2 - m_k$ planes, giving evidence of a scaling behaviour. The paths are well described by the moments of Brody's empirical distribution, which usually works very well for the intermediate statistics

$$P(s) = As^\beta \exp\left(-Bs^{\beta+1}\right) \tag{6.2}$$

where A and B are the normalization constants, and β is a free parameter. Note that for $\beta = 0$ the distribution is Poissonian, and for $\beta = 1$ it is Wigner. The scaling does not occur in the other obvious interpolation $tD + (1-t)G$, where D is random diagonal and G belongs to GOE [45].

In the investigation of the scaling properties of eigenvectors of BRM (§5), the scaling parameter was identified in the ratio b^2/N. Casati, Izrailev and Molinari [46] have shown that the whole $P(s)$ distribution for BRM only depends on the above scaling ratio. They used the empirical formula by Izrailev [33]

$$P(s) = As^\beta \exp\left[-\frac{\pi^2}{16}\beta s^2 - \frac{\pi}{2}\left(B - \frac{\beta}{2}\right)s\right] \tag{6.3}$$

which is more accurate than Brody's distribution in the description of the actual tail of the $P(s)$ distribution of Gaussian Ensembles. The spectral parameter β in (6.3), which fits the numerical P(s) distribution of BRM, was found to be close to the localization ratio β_1. As b^2/N increases, states range from exponentially localized to extended, and correspondingly, the level statistics changes from Poisson to Wigner-Dyson.

For fixed b/N and increasing N, Camarda [47] has shown that the BRM spectral statistics rapidly saturates to that of full random matrices, in agreement with the fact that $b^2/N \to \infty$.

The small spacing s behaviour of the distribution $P(s)$ of BRM was studied analytically by Molinari and Sokolov for the simplest possible case $n = 3$ and $b = 2$ [48], and extended by Grammaticos, Ramani and Caurier to larger tridiagonal random matrices [49,50]. The result, which has been proved analytically for $N = 3,4$ and numerically for the few next cases, states that $P(s)$ behaves as $s(-\log s)^{N-2}$ for small s.

The BRM ensemble so far discussed in its mathematical properties, is too simple to provide a justification for the intermediate statistics of physical systems, the main objection being the dependence of b on N, and the uncertain interpretation of the parameters. An answer is given by the Wigner ensemble.

7. Wigner Ensemble

Feingold, Leitner, Piro and Wilkinson have investigated the features of the matrix representation, in the semiclassical limit, of ergodic Hamiltonians [5.,31,52] and obtained a band structure, with increasing diagonal elements. They therefore introduced an ensemble of BRM matrices of size N, half-band b, and an additional diagonal part with increasing elements, with equal spacing α. A slightly different ensemble, with random elements taking values $\pm h$, was actually introduced by Wigner with the name of "bordered matrices" (see also [1]). For this reason we like to call the BRM ensemble with diagonal part, the Wigner Ensemble. In the tight binding picture, the diagonal part corresponds to the addition of a uniform electric field. Let us sketch their argument for the finiteness of the bandwidth. An operator $\hat{H} = \hat{H}_0 + \hat{H}_1$ has matrix representation $H_{ij} = E_i \delta_{ij} + \langle E_i | \hat{H}_1 | E_j \rangle$ in the basis of $\hat{H}_0$, ordered according to $E_i < E_{i+1}$. To measure the energy bandwidth of the i-th row they introduce the variance

$$(\delta E)_i^2 = \frac{\sum_j (E_i - E_j)^2 |H_{ij}|^2}{\sum_j' |H_{ij}|^2} = \frac{\langle E_i | [i\hat{H}_0, \hat{H}_1]^2 | E_i \rangle}{\langle E_i | \hat{H}^2 | E_i \rangle - \langle E_i | \hat{H} | E_i \rangle^2} \tag{7.1}$$

The expectation values can be computed as phase space integrals introducing the Wigner function $W_i(p,q)$ for the state $|E_i\rangle$ and the functions for the operators, for example according to the Weyl correspondence. The commutator then corresponds to a Moyal bracket. To lowest order in $\hbar$, these functions are the classical expressions of the Hamiltonians, and the Moyal bracket is a Poisson bracket multiplied by $\hbar^2$. For high quantum numbers, and in presence of classical ergodicity, phase space averages on Wigner's functions tend to microcanonical averages:

$$\int dp dq \, W_i(p,q) f(p,q) \to \{f\}_i = \int dp dq \frac{1}{\mu(E_i)} \delta(H_0(p,q) - E_i) f(p,q) \tag{7.2}$$

a result stated by Snirelman. With these approximations, the above energy width is

$$(\delta E)_i^2 \to \hbar^2 \frac{\{\{H_0, H_1\}^2\}_i}{\{H^2\}_i - \{H\}_i^2} \tag{7.3}$$

Assuming that these microcanonical averages exist, we end up with a finite band-width for the row i: $b_i \approx \delta E_i \rho(E_i)$, where $\rho(E)$ is the energy density of $\hat{H}_0$.

The presence of a uniform electric field enhances localization [53]. Consider for example the tight binding equation for a particle on a 1D lattice with a uniform field

$$\psi_{n+1} + \psi_{n-1} + \alpha n \psi_n = E \psi_n \qquad (7.4)$$

When $\alpha = 0$, it is solved by plane waves $\psi_n = \exp(\pm ipn)$, with $E = 2\cos p$, and the particle moves freely. For $\alpha \neq 0$ the equation is solved by Bessel functions, implying a rapid decay of the solution. The linear potential and the existence of an energy band do not allow the particle to propagate beyond an "electric length" measured by $\alpha \xi_{el} \approx 2$. For a lattice model described by BRM, the energy distribution is a semicircle of radius $\sqrt{b}$, and $\xi_{el} = \sqrt{b}/\alpha$. The other characteristic length is produced by the band structure, and is proportional to b^2. The ratio of the two lengths is the parameter $y = \alpha b^{3/2}$ introduced by Feingold et al. They investigated the level spacing distribution $P(s)$ of the Wigner ensemble for large N and various values of b and α, by fitting the numerical results with a Brody distribution (6.2). It resulted that Brody 's parameter β only depends on y; the distribution P(s) becomes Poissonian for $y \to 0$, and Wigner-Dyson for $y \to \infty$. This new scaling is physically more interesting than the one described in §6, as discussed also by Chirikov et al. [54], and make the Wigner Ensemble a candidate to describe the intermediate statistics.

The distribution of eigenvalues is studied in [55]. The localization length for the tight binding model of infinite length, modelled by infinite Wigner matrices, was discussed in [31] and [32] with the result

$$\xi_\infty(b, \alpha) = b^2 F(y) \qquad (7.5)$$

where $F(y)$ vanishes as $1/y$ for large y. The investigation of localization length in the Wigner ensemble for finite size N was carried by Feingold, Gioletta, Izrailev and Molinari [56]. They found numerically that the (entropy) length $\xi_N(b, \alpha)/N$ is a scaling function of the two parameters $x = b^2/N$ and y. For a review of the subject see also [57].

An attempt similar to Feingold's approach was undertaken by Prosen and Robnik [58] to design a matrix model for the spectral statistical properties of quantum Hamiltonian systems in the transition between chaos and integrability. Their motivation was to understand the effectiveness of Brody's distribution and the failure of Berry-Robnik formulae. The resulting ensemble, besides being banded with increasing diagonal elements, has the new property of sparseness.

References

1. M.L.Mehta: "Random Matrices", Academic Press (1991) 2nd Ed.
2. F.Haake: "Quantum Signatures of Chaos", Springer series in Synergetics vol. 54, Springer-Verlag (1991).
3. Bohigas: "Random matrix theories and chaotic dynamics", Les Houches session LII (1989) in Chaos and Quantum Physics, M.J. Giannoni, A.Voros and J.Zinn-Justin eds. Elsevier Science Publ. 1991.

4. H.A.Weidenmüller: "Stochastic Scattering theory or Random Matrix models for fluctuations in Microscopic and Mesoscopic Systems", in "Chaos and Quantum Chaos", ed. W.D.Heiss, Lect. Notes in Phys. 411, Springer-Verlag (1992).
5. D.Poilblanc, T.Ziman, J.Bellissard, F.Mila and G.Montambaux, Europhys. Lett. 22 (1993) 537.
6. E.P.Wigner, Ann. Math. 62 (1955) 548 and 65 (1957) 203.
7. F.M.Izrailev, Phys. Rep. 196 (1990) 299.
8. D.L.Shepelyansky, Phys. Rev. Lett. 56 (1986) 677.
9. S.Fishman, R.E.Prange and M.Griniasty, Phys. Rev. A39 (1989) 1628.
10. G.Casati and L.Molinari, Progr. Theor. Phys. Suppl. 98 (1989) 287.
11. S.Fishman, D.R.Grempel and R.E.Prange, Phys. Rev. Lett. 49 (1982) 509.
12. G.Casati, I.Guarneri, F.Izrailev and R.Scharf, Phys Rev. Lett. 64 (1990) 5.
13. E.Doron and S.Fishman, Phys. Rev. Lett. 60 (1988) 867.
14. G.Casati, I.Guarneri and D.L.Shepelyansky, Phys. Rev. Lett. 62 (1989) 345.
15. P.W.Anderson, Phys. Rev. 109 (1958) 1492.
16. P.Lloyd, J. Phys. C2 (1969) 1717.
17. S.Fishman: "Quantum Localization", Proc. Int. School "Enrico Fermi" on Quantum Chaos, Varenna 1991. G.Casati, I.Guarneri and U. Smilansky Eds.
18. H.L.Cycon, R.G.Froese, W.Kirsch and B.Simon: "Schrödinger Operators", Springer-Verlag, 1987.
19. M.P.Sörensen and T.Schneider, Z.Phys. B82 (1991) 115.
20. P.A.Lee and T.V.Ramakrishnan, Rev. Mod. Phys. 57 (1985) 287.
21. J.L.Pichard, J. Phys. C19 (1986) 1519.
22. K.Życzkowski, L.Molinari and F.M.Izrailev: "Level curvature and metal-insulator transition in 3D Anderson model", (1994) submitted to J. Physique I.
23. B.I.Shklovskii, B.Shapiro, B.R.Sears, P.Lambrianides and H.B.Shore, Phys. Rev. B47 (1993) 11487.
24. F.J.Dyson, Phys. Rev. 92 (1953) 1331.
25. M.Feingold, Europhys. Lett. 17 (1992) 97.
26. M.Kus, M.Lewenstein and F.Haake, Phys. Rev. A44 (1991) 2800.
27. S.A.Molchanov, L.A.Pastur and A.M.Khorunzhii, Theor. and Math. Phys. 90 (1992) 108.
28. G.Casati and V.Girko: Random Oper. and Stoch. Equ. 1 (1993) 15.
29. G.Casati, I.Guarneri, F.Izrailev, L.Molinari and K.Życzkowski: "Band Random Matrices, Curvature and Conductance in Disordered Media", (1994) submitted to Phys. Rev. Lett.
30. J.Zakrzewski and D.Delande, Phys. Rev. E47 (1993) 1650.
31. M.Wilkinson, M.Feingold and D.M.Leitner, J. Phys. A24 (1991) 175.
32. Y.V.Fyodorov and A.D.Mirlin, Phys. Rev. Lett. 67 (1991) 2405.
33. F.M.Izrailev, Phys. Lett. A134 (1988) 13.
34. G.Casati, F.Izrailev and L.Molinari, Phys. Rev. Lett. 64 (1990) 1851.
35. S.N.Evangelou and E.Economou, Phys. Lett. A151 (1990) 345.
36. Y.V.Fyodorov and A.D.Mirlin, Phys. Rev. Lett. 69 (1992) 1093.
37. Y.V.Fyodorov and A.D.Mirlin, Phys. Rev. Lett. 71 (1993) 412.
38. K.Życzkowski, M.Lewenstein, M.Kus and F.Haake, Phys. Rev. A45 (1992) 811.
39. Y.V.Fyodorov and A.D.Mirlin, J. Phys. A26 (1993) L551.
40. G.Casati, I.Guarneri, F.Izrailev, L.Molinari and S.Fishman: J. Phys.: Condens. Matter 4 (1992) 149.
41. L.Molinari, J. Phys.: Condens. Matter 5 (1993) L319.
42. T.H.Seligman, J.J.M.Verbaarschot and M.R.Zirnbauer, Phys. Rev. Lett. 53 (1984) 215.
43. T.H.Seligman, J.J.M.Verbaarschot and M.R.Zirnbauer, J. Phys. A18 (1985) 2751.
44. T.Cheon, Phys. Rev. Lett. 65 (1990) 529.
45. T.Cheon, Phys. Rev. A42 (1990) 6227.
46. G.Casati, F.Izrailev and L.Molinari, J. Phys. A24 (1991) 4755.
47. H.S.Camarda, Phys. Rev. A45 (1992) 579.
48. L.Molinari and V.V.Sokolov, J. Phys. A22 (1989) L999.
49. B.Grammaticos, A.Ramani and E.Caurier, J.Phys. A23 (1990) 5855.
50. E.Caurier, A.Ramani and B.Grammaticos, J. Phys. A26 (1993) 3845.
51. M.Feingold, D.M.Leitner and O.Piro, Phys. Rev. A39 (1989) 6507.
52. M.Feingold, D.M.Leitner and M.Wilkinson, Phys. Rev. Lett. 66 (1991) 986.

53. M.Luban and J.Luscombe, Phys. Rev. **B34** (1986) 3674.
54. G.Casati, B.V.Chirikov, I.Guarneri and F.M.Izrailev, Phys. Rev. **E48** (1993) R1613.
55. D.M.Leitner and M.Feingold, J. Phys. **A26** (1993) 7367.
56. M.Feingold, A.Gioletta, F.M.Izrailev and L.Molinari, Phys. Rev. Lett. **70** (1993) 2936.
57. M.Feingold and D.Leitner: "Semiclassical localization in time independent K-systems", Proc. of Mesoscopic Systems and Chaos, a Novel Approach, Trieste 1993.
58. T.Prosen and M.Robnik, J.Phys. **A26** (1993) 1105.

ROBUST SCARRED STATES

D. RICHARDS
Mathematics Faculty, Open University,
Milton Keynes, MK7 6AA, England.

Abstract. For periodically forced systems we show that the scarred state corresponding to the unstable periodic orbit of the main resonance has a lower ionisation rate than adjacent states because a representation exists in which coupling to other states is minimal. Our results explains why this scarred state has been observed in recent experiments on the microwave ionisation of excited hydrogen atoms; other previously unexplained features are also elucidated.

1. Introduction

A quantal state of a classically chaotic system has a *scar* of a periodic orbit if its density near this orbit is significantly larger than that of the statistically expected density, Heller (1991). Such states have been observed in numerous numerical calculations Wintgen and Hönig (1989), Eckhardt *et al* (1989), Waterland *et al* (1988) and Provost and Baranger (1993), and their effect on the microwave ionisation of excited hydrogen atoms is seen by the enhanced stability of isolated states Koch (1992), Sauer *et al* (1992), Sirko *et al* (1993), Jensen *et al* (1989). The essential property of a scarred state is that it is a quantal feature built upon a classical phase space structure, generally an unstable periodic orbit, which has a vanishingly small effect on the behaviour of an ensemble of orbits but a noticeable affect on the equivalent wave function. Such states are important because classically chaotic systems possess infinitely many unstable periodic orbits and the associated scars produce quantal effects which disappear rather slowly as the classical limit, $\hbar \to 0$, is approached, Wintgen and Hönig (1989).

In this paper we consider the scarred states of periodically driven one-dimensional hydrogen atom, although the analysis we present is applicable to any driven one-dimensional system. We are not concerned with *what* these states are but *why* they ionise less readily than adjacent states.

Jensen *et al* (1989) have shown numerically that a scarred state of the 1*d* hydrogen atom in a periodic electric field resides on an unstable periodic orbit associated with a resonant island: this result suggests that it is essential to treat the resonant island in its entirety and not the two constituent fixed points, corresponding to the stable and unstable periodic orbits, separately. Whilst these numerical results show *what* these scarred states are they do not explain *why* they ionise less readily.

Here we provide an alternative analysis which shows why this state is more stable than adjacent states; although we deal only with time-dependent, resonantly forced one-dimensional systems, the same type of analysis may be appropriate to two-dimensional conservative systems as the structures in the surface of section plots are similar because they are produced by the same resonant dynamics. We consider

161

Z. Haba et al. (eds.), Stochasticity and Quantum Chaos, 161–176.

only the periodically driven $1d$ hydrogen atom, but it will be clear that the method is quite general and can be applied to any periodically driven $1d$ system. The fields considered are weak in the sense that between 10% and 50% of the atoms are ionised during interaction times of at least a hundred field periods, so the ionisation rate per field period is small and almost all ionisation is from states far removed from the initial state, see Leopold and Richards (1991). This means that the features in the ionisation probability that we wish to understand are determined by the bound-state dynamics. Thus in the theory of Section 2 we ignore the continuum, but later we show, numerically, that the predictions of this theory are consistent with calculations including the continuum.

We shall show that a quantal state associated with the unstable periodic orbit of resonant islands is more weakly coupled to higher lying states, from which ionisation occurs, than adjacent states. The reason for this is both simple and bizarre; it is because the classical motion on the associated separatrix is slow but the classical and quantal consequences of this are quite different. Classically this slow motion is significantly affected by the faster terms of the Hamiltonian and this leads to a narrow chaotic band around the separatrix. Quantally, these different time scales result in the faster terms having less effect, so that the state associated with the separatrix state is more weakly coupled to the continuum than adjacent states. In order to demonstrate this physical effect mathematically it is necessary to find a transformation which highlights the slow motion on the separatrix; this is constructed in section 2.3, after some necessary preamble.

2. Theory

2.1. INTRODUCTION

The scars of interest here are related to the main resonance of the driven system, produced when the driving frequency Ω is close to the classical orbital frequency, $\omega(n_0)$ where n_0 is the initial level. Other scars almost certainly exist; experimental measurements of threshold fields suggest that scarred states exist when $\Omega_0 = \Omega/\omega(n_0) \simeq 1/q$, with $q = 1, 2, \ldots$: the present analysis shows that, for sufficiently large quantum numbers, scarred states will be observable if $\Omega_0 \simeq p$, $p = 1, 2, \cdots$, and suggests that they will also be observable near any rational frequency ratio.

In order to concentrate the reader's mind we compare, in figure 1, the experimentally observed scars with classical Monte Carlo simulations. In these experiments the wave guide frequency is fixed at 30.36 GHz so the scaled frequency is given by $\Omega_0 = 30.36(0.00533757n_0)^3$, with the initial principal quantum number in the range $55 \leq n_0 \leq 72$. This graph shows the laboratory field required to produce 10% ionisation as a function of the scaled frequency Ω_0; the local maximum at $\Omega_0 \simeq 1.3$ has been shown by Jensen et al (1989), to be a consequence of the system being in a state whose wave function resides mainly on an unstable orbit: notice that the classical thresholds show no structure near this frequency. In a series of detailed experiments Sauer et al (1992) have shown that this local maximum in $F_0(10)$ exists at the same scaled frequency but for different values of Ω and n_0; that is, they have shown empirically that the position of this maximum scales classically, even though

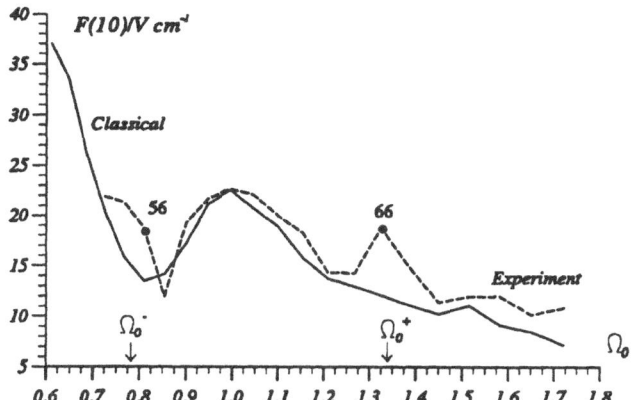

Fig. 1.   Graph showing the 10% threshold field as a function of the scaled frequency, equation 4. The solid line depicts the experimental results the dashed line the classical Monte Carlo simulations of the $3d$ Hamiltonian; these results are taken from Sirko *et al* (1993). The separatrix state, defined in section 2.2, see also equation 26, with $F_s = 0$, are the marked levels $n_0 = 56$ and 66; note that the threshold fields for each of these are approximately the same.

it is not an observable feature of the classical thresholds.

The experiments which produce such scars perturb the atom by an electric field with a potential of the form $Fz\lambda(t)\cos(\Omega t+\delta)$, where the phase δ is averaged over on present experiments: the field envelope, $\lambda(t)$, varies slowly by comparison to $\cos\Omega t$ and $\lambda(t) = 0$ for $t \leq 0$ and $t \geq T_m \gg 2\pi/\Omega$. Sometimes, but not always, $\lambda(t) = 1$ for many field periods; the existence of the scars seems to be independent of the form of $\lambda(t)$, provided it varies sufficiently slowly, see Sauer *et al* (1992). For a theoretical analysis it is most convenient to deal with the case where $\lambda(t)$ rises slowly to unity over a period $0 < t < T_a$, remains constant for a time $T_m - 2T_a \gg 2\pi/\Omega$ and then falls slowly to zero at $t = T_m$. Then most ionisation occurs during the period when λ is constant via decay from excited states, Leopold and Richards (1991): because only a small proportion of the atoms ionise during the interaction the ionisation rate per field period must be small hence the ionisation during $(0, T_a)$ may be ignored. Thus with this pulse shape we may consider the switch-on time, $0 < t < T_a$, as a period during which the system is prepared and the problem is conveniently divided into two parts: the dynamics of the switch-on and the properties of the solution of the time-dependent Schrödinger equation when $\lambda = 1$.

We consider the switch-on dynamics in section 3 and concentrate on the dynamics in constant amplitude fields in the present section. The aim of the ensuing analysis is not to solve Schrödinger's equation, but to explain why the scar shown in figures 1 and 4 exist, in other words why the state associated with the unstable periodic orbits is more stable than adjacent states. We do this by showing that there is a representation, (Q, P), in which the Hamiltonian can be written in the form $K_0(Q, P) + F K_1(Q, P, F)$ and where a particular eigenstate of K_0, associated with the unstable periodic orbit in the original representation, is perturbed less by $F K_1$ than adjacent states, so the equivalent Floquet states have smaller contribution from excited states and hence smaller decay rates. We show that these states coincide with the scar shown in figure 1 and, in section 4, with scars obtained by the direct numerical solution of the the time-dependent Schrödinger equation for a one-dimensional hydrogen atom.

In our analysis we use the dipole form of the Hamiltonian of a one-dimensional hydrogen atom moving through a periodic electric field fixed in the laboratory, Leopold and Richards (1991),

$$\mathcal{H} = \frac{1}{2\mu} p^2 - \frac{e^2}{z} - F\lambda(t) z \cos \Omega t, \quad z \geq 0. \tag{1}$$

The analysis we present, however, can be applied to any periodically forced one-dimensional system; in section 4 we shall modify this Hamiltonian with the addition of a static electric field, and there show that the following results remain true for a continually varying field envelope.

2.2. Theory for constant amplitude fields

In this section we consider the case $\lambda(t) = 1$, for all time, as we need to understand the solutions of Schrödinger's equation near a resonant perturbation. These solutions are most easily understood using classical dynamics and the unperturbed angle-action variables, (θ, I),

$$z = \frac{2I^2}{\mu e^2} \sin^2 \psi, \quad p = \frac{\mu e^2}{I \tan \psi}, \quad \theta = \omega t + \delta = 2\psi - \sin 2\psi,$$

where $\omega(I) = \mu e^4 I^{-3}$ is the Kepler frequency. On expanding $\sin^2 \psi$ as a Fourier series in θ, we obtain, after some analysis and on ignoring all rapidly varying terms,

$$\mathcal{H} = -\frac{\mu e^4}{2I^2} + \frac{F I^2}{\mu e^2} \sum_{s=1}^{\infty} \frac{J_s'(s)}{s} \cos(s\theta - \Omega t), \tag{2}$$

where $J_s(x)$ is an ordinary Bessel function. Near the main resonance $I \simeq I_r$, the resonant action I_r being defined below, it is convenient to remove the slowly varying term $\cos(\theta - \Omega t)$, using the time-dependent canonical transformation $\phi = \theta - \Omega t$. It is also convenient to use scaled units, where the scaling is relative to the resonant orbit,

$$I_r = \left(\frac{\mu e^4}{\Omega} \right)^{1/3}, \quad F_r = \frac{F I_r^4}{\mu^2 e^6}, \quad I_r = n_r \hbar. \tag{3}$$

Typically $F_r \sim 0.05$. In terms of the usual scaled variables, F_0 and Ω_0, relative to the initial state, n_0, we have,

$$F_0 = F_r \left(\frac{n_0}{n_r} \right)^4 \quad \text{and} \quad \Omega_0 = \left(\frac{n_0}{n_r} \right)^3. \tag{4}$$

In this moving reference frame the new Hamiltonian is conveniently wr.tten in the form $H = H_0 + H_1$ where

$$H_0(\phi, I) = \Omega \left\{ -\frac{I_r^3}{2I^2} - I + \frac{\alpha F_r}{I_r} I^2 \cos\phi \right\}, \tag{5}$$

$$H_1(\phi, I) = I^2 \Omega \frac{F_r}{I_r} \sum_{s=2}^{\infty} \frac{J_s'(s)}{s} \cos(s\phi + (s-1)\Omega t), \tag{6}$$

where $\alpha = J_1'(1) \simeq 0.325$.

Since we are interested in motion near the resonant island we write $I = I_r + X$ and it is sufficiently accurate to expand to $O(X^2)$ to obtain the simpler approximation

$$H_0(\phi, X) = -\Omega \left\{ \frac{3X^2}{2I_r} - \alpha F_r I_r \cos\phi \right\}. \tag{7}$$

Near a resonance any one-dimensional system can be written in the form of the 'pendulum' Hamiltonian, H_0, and a periodic perturbation H_1; the quantal equivalent of this analysis is more complicated, but gives essentially the same result, and is given by Zaslavsky (1981), see also section 3. In this form the 'perturbation' H_1 is not particularly small and has magnitude comparable to $\alpha I_r F_r$ so neither classical nor quantal perturbation theory are very useful. Moreover, the form of the perturbation H_1 provides no clue about how it will affect the different eigenstates of H_0. The unperturbed motion due to H_0, however, suggests that H_1 has rather less effect than its relative magnitude would suggest. The following properties of the unperturbed motion are important.

(a) There are two types of unperturbed motion: rotational motion for energies $E < E_s = -\alpha F_r I_r \Omega$ and librational motion for the energies $E_s < E < -E_s$, with the separatrix of energy E_s dividing the phase space into invariant regions. There is no motion for $E > -E_s$.

(b) There is a stable fixed point at $\phi = X = 0$, associated with a stable periodic orbit in the original representation. Near this fixed point the motion is close to that of a linear oscillator with natural frequency $\omega_L = \Omega\sqrt{3\alpha F_r}$, which is typically of order 0.2Ω. For a variety of reasons, which will be discussed later, the lowest energy states in the island are affected less by H_1 than the other states, so the associated Floquet states are approximately the same as the eigenstates of H_0.

(c) There is an unstable fixed point of H_0 at $\phi = \pm\pi$, $X = 0$ associated with an unstable periodic orbit in the original representation. Near the separatrix the period of the classical motion is large and tends to infinity as $E \to E_s$; we shall show that near E_s the perturbation H_1 has qualitatively different effects on the classical and quantal motion: classically, for any $F > 0$, the perturbation produces chaotic motion, while the quantal states are relatively well behaved and locally less sensitive to H_1. The states with energy close to E_s we loosely call the *separatrix states*; one of these is less perturbed by H_1 than adjacent states. In the following subsection we explain this phenomenon and show how it is responsible for the maximum at $\Omega_0 \simeq 1.3$ in Figure 1.

(d) The area inside the separatrix is proportional to the number of librational states,

$$N_l \simeq 0.838 n_r \sqrt{F_r}. \tag{8}$$

For typical ionising field this is small, about 15, despite the large initial principle quantum number, typically 50–100. Close to the separatrix quantal effects are important so there is no unambiguous connection between classical orbits and quantal states. For this reason it is difficult to define a priori which separatrix state is least affected by H_1.

2.3. REMOVAL OF RAPIDLY VARYING TERMS: GENERAL THEORY

Near the separatrix of H_0 the period, T, of the motion is long so the perturbation H_1 is relatively rapidly varying; thus we should expect it to have small effect even though it has the same order of magnitude as the $\cos\phi$ term of H_0, equation 7. Here we show that there exists a representation in which the dominant term of the perturbation is proportional to the square of the momentum, the mean of which is smallest on the separatrix.

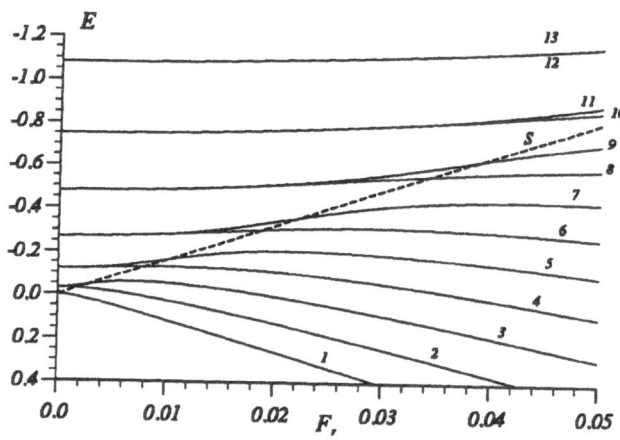

Fig. 2.   Diagram showing the energy levels of the resonant Hamiltonian, equation 18, as a function of the scaled field F_r, in the case $I_r = 50\hbar$. The separatrix is the dashed line S: below S the classical motion is all librational and above S it is rotational.

Rapidly oscillating perturbations are normally encountered in Hamiltonians in which the perturbation is linear in the generalised coordinate; then the acceleration gauge or, equivalently the Kramers-Hennenberger transformation, Kramers (1956) and Henneberger (1968), can be used to obtain a representation in which the Hamiltonian is $O(\Omega^{-2})$. This transformation is frequently used in quantal calculations, see for instance Reed and Burnett (1990) and Burnett *et al* (1992), but it is important to notice that because the original perturbation is linear in the generalised coordinate the quantal operator equivalent of the new Hamiltonian can be obtained *exactly* from the classical Hamiltonian.

In the more general case the high frequency term is not linear; a similar classical analysis is possible although it is rather more complicated and it is not obvious what form the exact quantal equivalent takes, although the approximate semiclassical form of the transformed Hamiltonian is easily obtained. Our starting Hamiltonian,

$$H(q, p, \tau) = -\frac{3p^2}{2I_r} + \alpha I_r F_r \cos q + F_r I_r \sum_{s=2}^{\infty} \frac{J_s'(s)}{s} \cos(sq + (s-1)\tau),$$

is obtained from equation 7, setting $I = I_r$ in equation 6 and on defining the new time $\tau = \Omega t$. Whenever the unperturbed motion changes little during one oscillation of the perturbation we may, for short times, ignore the $\cos q$ term to obtain the approximate motion

$$q = Q + 3F_r \sum_{s=2}^{\infty} \frac{J_s'(s)}{(s-1)^2} \sin \Theta_s - \frac{3P\tau}{I_r}, \tag{9}$$

$$p = P - F_r I_r \sum_{s=2}^{\infty} \frac{J_s'(s)}{s-1} \cos \Theta_s, \quad \text{where} \quad \Theta_s = sQ + (s-1)\tau. \tag{10}$$

This suggests that a transformation $(q, p) \rightarrow (Q, P)$ obtained by ignoring the last term of 9 would produce simpler equations of motion; this transformation, however, is not canonical, but it may be made so by using equation 10 to suggest an extra term. The details of this derivation are given in Leopold and Richards (1993); the final generating function is

$$F_3(Q, p, \tau) = -p \left\{ Q + 3F_r \sum_{s=2}^{\infty} \frac{J_s'(s)}{(s-1)^2} \sin \Theta_s \right\} - F_r I_r \sum_{s=2}^{\infty} \frac{J_s'(s)}{s(s-1)} \sin \Theta_s \tag{11}$$

which gives the canonical transformation

$$q = Q + g(Q, \tau), \quad p = \frac{1}{f(Q, \tau)} \left\{ P - F_r I_r \sum_{s=2}^{\infty} \frac{J_s'(s)}{s-1} \cos \Theta_s \right\} \tag{12}$$

where

$$f(Q, \tau) = 1 + 3F_r \sum_{s=2}^{\infty} \frac{sJ_s'(s)}{(s-1)^2} \cos \Theta_s \quad \text{and} \quad g(Q, \tau) = 3F_r \sum_{s=2}^{\infty} \frac{J_s'(s)}{(s-1)^2} \sin \Theta_s.$$

The transformation looks more complicated than it really is because most terms are relatively small. Using this transformation the new Hamiltonian is

$$K(Q, P, \tau) = -\frac{3P^2}{2I_r} + \alpha I_r F_r \cos Q + \sum_{k=1}^{4} K_k(Q, P, \tau) \tag{13}$$

where

$$K_1(Q, P, \tau) = \frac{3P^2}{2I_r} \left(1 - \frac{1}{f(Q, \tau)^2} \right) \tag{14}$$

$$K_2(Q, P, \tau) = F_r I_r \sum_{s=1}^{\infty} \frac{J_s'(s)}{s} \left\{ \cos[\Theta_s + sg(Q, \tau)] - \cos \Theta_s \right\} \tag{15}$$

$$K_3(Q,P,\tau) = \frac{3I_r F_r^2}{f(Q,\tau)}\left(1 - \frac{1}{2f(Q,\tau)}\right)\left[\sum_{s=2}^{\infty}\frac{J_s'(s)}{s-1}\cos\Theta_s\right]^2 \qquad (16)$$

$$K_4(Q,P,\tau) = -\frac{3F_r P}{f(Q,\tau)^2}(f(Q,\tau)-1)\sum_{s=2}^{\infty}\frac{J_s'(s)}{s-1}\cos\Theta_s. \qquad (17)$$

This rather complicated Hamiltonian is not suited for computation but helps understand the effect of the perturbation; in particular the magnitude and form of the perturbations $K_1, \cdots, K_4$ show why one of the separatrix states is relatively more stable than adjacent states.

2.4. RELATIVE STABILITY OF THE SEPARATRIX STATE

The advantage of this new Hamiltonian over the original becomes apparent when we consider the magnitude of the perturbations and the manner in which they vary along the unperturbed motion.

An estimate of the sizes of these perturbations is obtained by expanding in powers of F_r and computing the root-mean-square over Q and τ; we find

$$K_1 \sim 3.0\frac{P^2 F_r}{I_r}, \quad K_2 \sim 0.14 I_r F_r^2, \quad K_3 \sim 0.05 I_r F_r^2 \quad \text{and} \quad K_4 \sim 0.5 P F_r^2.$$

Typically $F_r \sim 0.05$, $I_r \sim 50\hbar$, giving $N_l \sim 10$; near the separatrix $P \sim N_l\hbar/2 \sim 5\hbar$ so

$$K_1 \sim 0.08\hbar, \quad K_2 \sim 0.02\hbar, \quad K_3 \sim 0.006\hbar \quad \text{and} \quad K_4 \sim 0.007\hbar.$$

From this we see that K_1 is the dominant perturbation, the remaining terms being relatively small. In the original representation, equations 5 and 6 the mean of the perturbation is $0.25 I_r F_r \sim 0.6\hbar$. Thus, the perturbation in this representation relatively small; but of more significance is the fact that the dominant perturbation, K_1, is proportional to P^2 whose mean has a local minimum on the separatrix, see table I below; to understand why this happens we need to discuss the quantum mechanics of this system.

An approximation to the quantal equations of motion is obtained from the Hamiltonian $K(Q,P,\tau)$ by defining the operator $\hat{P} = -i\hbar\partial/\partial Q$ and symmetrising the products of powers of P and functions of Q wherever necessary. Using this procedure and any basis set we can construct an approximation to the time-dependent Schrödinger equation which differs from the exact equation by terms $O(\hbar^2)$.

Because the perturbations are small the natural basis are the eigenfunctions of the resonant Hamiltonian

$$K_0(Q,P) = -\frac{3P^2}{2I_r} + \alpha I_r F_r \cos Q. \qquad (18)$$

Thus, Schrödinger's equation can be written in the form of Mathieu's equation, Abramowitz and Stegun (1965, page 722).

For $F_r > 0$ the eigenfunctions are labeled by an index $k = 0, 1, 2, \cdots$. For $F_r = 0$ the eigenfunctions are $\exp(imQ)/\sqrt{2\pi}$, $m = 0, \pm 1, \pm 2, \cdots$. Later, in section 3, we

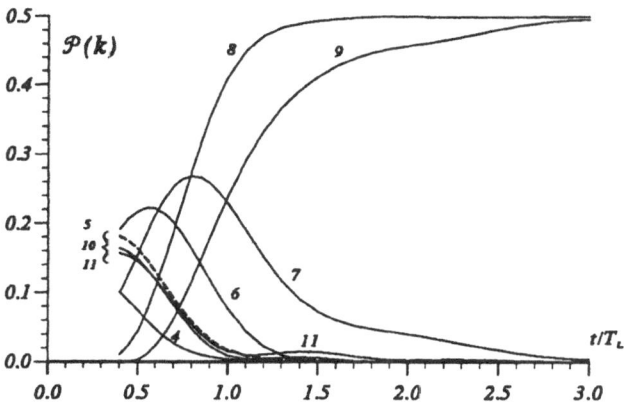

Fig. 3. Graph showing the probabilities $\mathcal{P} = |\langle k:1|m\rangle|^2$ for the initial state $m = 4$ and various values of k as a function of the switch time. As in Figure 3 $I_r = 50\hbar$ and $F_r = 0.04$.

shall be interested in the adiabatic connection between the $F_r = 0$ and the $F_r > 0$ eigenfunctions so, for later reference, we note that

$$m = 0 \to k = 1, \quad m = \pm 1 \to k = 2 \text{ and } 3, \quad \text{etc.}$$

This is shown schematically in figure 2 where the energy levels of this system are plotted as a function of F_r, for $I_r = 50\hbar$, together with the separatrix energy $-E_s = \alpha I_r F_r$, shown by the dashed line; the region below the dashed line corresponds to librational motion and above to rotational motion. When $F_r = 0$, $E \leq 0$, and all motion is rotational and there is a degeneracy corresponding to classical motion with $\dot{Q} > 0$ and $\dot{Q} < 0$ having the same energy. For $F_r > 0$ this degeneracy is lifted but not significantly until the energy approaches E_s.

The classical value of the mean square momentum $\langle P^2 \rangle_{cl}$ is zero on the separatrix. As the energy decreases from its highest value at the island centre, $\alpha I_r F_r$, the mean increases, reaches a maximum before decreasing to zero on the separatrix: for the rotational motion $\langle P^2 \rangle_{cl}$ increases monotonically with decreasing energy.

The quantal mean behaves similarly, but it clearly cannot be zero near the separatrix. In table I we show values of $\langle P^2 \rangle_{qu}$ for the case $n_r = 50$ and for various values of F_r: it is seen that the mean has a distinct minima, shown in italics, at a state close to the separatrix: it is instructive to compare the numerical values in this table with those in table II.

At the beginning of this section we showed that K_1 was the dominant perturbation: because this is proportional to P^2 we expect it to have less effect on the separatrix state than on adjacent states. In this representation the perturbation is relatively small so an estimate of the magnitude of the effect of the perturbations on the eigenfunctions of K_0 can be obtained using first-order perturbation theory to

TABLE I
Values of the mean momentum squared, $\langle P^2 \rangle_{qu}$, for particular eigenstates of K_0, $I_r = 50\hbar$ and for various values of the scaled field. We
also show the approximate number of states in the island, as computed from equation 8, and it is seen that in the vicinity of the
separatrix, $k \simeq N_l$, $\langle P^2 \rangle_{qu}$ has a local minimum.

F_r	N_l	$k = 3$	4	5	6	7	8	9	10
0.01	4.2	3.0	4.3	2.9	8.6	8.4	15.8		
0.02	5.9	4.9	6.3	6.5	8.8	6.1	15.0	14.8	
0.03	7.3	6.2	8.2	9.6	10.8	8.8	14.6	11.7	23.5
0.04	8.4	7.4	9.6	11.8	13.2	13.7	15.8	10.4	22.6
0.05	9.4	8.3	11.1	13.6	15.6	17.0	18.2	14.8	22.3

compute the total transition probability out of a given state during a field period:

$$\mathcal{P}(k) = \sum_n \left| \frac{i}{\hbar} \int_0^{2\pi/\Omega} dt \, \langle k|V|n \rangle \exp i(E_k - E_n)t/\hbar \right|^2 \tag{19}$$

where $\langle Q|k \rangle$ and E_k are the eigenfunctions and values of $K_0(Q, P)$, and $V = K_1 + K_2 + K_3 + K_4$ is the perturbation given in equation 13.

In table II we show values of this transition probability for $n_r = 50$: it is seen that for each F_r it has a minimum coinciding with the minimum of $\langle P^2 \rangle_{qu}$ shown in table I.

TABLE II
Table showing $100\mathcal{P}(k)$ for the same data as table I; it is seen that in the
vicinity of the separatrix, $k \simeq N_l$, $\mathcal{P}(k)$ has a local minimum.

F_r	N_l	$k = 3$	4	5	6	7	8	9	10
0.01	4.2	0.013	0.020	0.013	0.050	0.048	0.148	0.148	
0.02	5.9		0.197	0.222	0.317	0.221	0.677	0.622	1.594
0.03	7.3		0.770	1.000	1.230	1.091	1.880	1.499	3.789
0.04	8.4				3.352	3.744	4.522	3.193	7.255
0.05	9.4				7.187	8.485	9.649	8.384	12.619

Since ionisation occurs predominantly from states outside the island the transition probabilities to these states, obtained as in equation 19 but by limiting the range
of the sum, have also been computed to check that they have the same behaviour

In order to compute these probabilities it is necessary to truncate the infinite sums
in $K_k(Q, P, \tau)$, equations 14–17. The values shown were obtained by truncating all

series at $s = 4$; truncating at $s = 2$ gave probabilities having the same behaviour and similar values.

To summarise, we have constructed a new representation, defined by the generating function of equation 11, which removes the unimportant oscillations produced by the perturbation. In this representation the magnitude of the perturbation is smaller and has a minimum on the separatrix state. The use of first-order perturbation theory shows that this separatrix state is least affected by the the ignored terms of the Hamiltonian, so has smaller overlap with the states of higher quantum numbers from which ionisation occurs.

3. Preparation of the separatrix states

In the previous section we ignored the envelope and dealt only with a constant amplitude field: we showed that the separatrix state is locally more stable than adjacent states. Here we consider how the separatrix state is prepared by the slow switch-on of the field. We consider only the unperturbed Hamiltonian, equation 18, but add a switch to the coefficient of $\cos Q$ so it becomes

$$K(Q, P : \lambda) = -\frac{3P^2}{2I_r} + \alpha F_r I_r \lambda(t) \cos Q, \quad 0 \le \lambda \le 1, \tag{20}$$

with $\lambda(t) = 0$ for $t < 0$ and unity for $t > T_a$.

If λ is a non-zero constant the eigenfunctions, $\langle Q | k : \lambda \rangle$, and eigenvalues, $E_k(\lambda)$, $k = 0, 1, \cdots$, depend upon λ. The eigenfunctions of $K(Q, P : 0)$ are $\langle Q | m \rangle = \exp(imQ)/\sqrt{2\pi}$, $m = 0, \pm 1, \pm 2 \cdots$. We require a condition on T_a such that the probability $|\langle k : 1 | m \rangle|^2$ differs from unity or one half, see below, by some specified small amount. Classically we require that a rotational phase curve, $\mathcal{C}(0)$, of the initial Hamiltonian, $P = P_0 =$ constant, is transformed into a curve $\mathcal{C}(T_a)$ lying close to the librational phase curves of $K(Q, P : 1)$ with action equal to $P_0/2$.

For a general one-dimensional system if λ changes sufficiently slowly classical phase curves and quantal wave functions adjust adiabatically so that, to a very good approximation $|\langle k : \lambda_1 | k : \lambda_2 \rangle|^2 = 1$ and $\mathcal{C}(T_a)$ is everywhere close to the initial phase curve of the same action. In the present problem the switch introduces a separatrix into phase space. For the classical system this means that the principle of adiabatic invariance, see Percival and Richards (1982, chapter 9), no longer holds throughout the switch-on period for initial actions $|P_0| < N_L \hbar/2$ as at some time the instantaneous frequency is zero; nevertheless for sufficiently long switched most final actions are close to $|P_0|/2$ provided $|P_0| < N_L \hbar/2$, see Figure 3. The quantal system is not so severely affected by the separatrix as, in this example, there are no small energy gaps, because the degenerate states are not coupled.

Thus there ought to be a qualitative similarity between the classical and quantum dynamics, although the details will differ. Here we consider only the quantal system; the comparison between the classical and quantum dynamics is given in Leopold and Richards (1993)

The naïve quantisation of equation 20 is incorrect as $I_r/\hbar$ is generally not an integer. Thus we need to start from equation 5, define $P = I - \tilde{I}_r$, where $\tilde{I}_r/\hbar$ is the nearest integer to $I_r/\hbar$, so apart from an irrelevant constant the equivalent of equation 18 is

$$K_0(Q, P; \lambda) = -\frac{3I_r^3}{2\tilde{I}_r^4}(P + \beta)^2 + \alpha F_r \frac{\tilde{I}_r^2}{I_r}\lambda(t)\cos Q \qquad (21)$$

where $\beta = \tilde{I}_r((\tilde{I}_r/I_r)^3 - 1)/3 \simeq \tilde{I}_r - I_r$. We can now quantise this system by defining the momentum operator $P = -i\hbar\partial/\partial Q$; the eigenfunctions for λ =constant are thus solutions of Mathieu's equation, satisfying the boundary conditions

$$f(v) = f(v + \pi)\exp(-2i\pi\beta/\hbar), \quad v = Q + \pi/2.$$

When λ varies we solve the time-dependent Schrödinger equation using a basis of unperturbed states $\langle Q|m\rangle$ one of which is the initial state. In all these calculations we took $\lambda(t) = x^2(2 - x^2)$, $x = t/T_a$ with $0 \leq t \leq T_a$.

In the special case where $I_r/\hbar$ is an integer, $\beta = 0$, the eigenfunctions are Mathieu functions of period π and are alternately odd and even so most initial states are switched adiabatically into a pair of eigenfunctions. The exception is the 'resonant' state, $m = 0$, which is even and switched adiabatically into the ground state of the island, $k = 1$; this is shown in detail by Leopold and Richards (1993).

In order to populate the separatrix state we need to start in an initial state $m \simeq \pm N_L/2$. For instance if $F_0 = 0.04$ and $n_r = 50$ then $N_L \simeq 8.4$, so from Figure 2, we should expect to populate the separatrix states, $k = 8$, 9 by starting in the rotational state $m = 4$. In Figure 3 we show the variation of the final population with the switch-time. In this case the required states are populated when $T_a \simeq 3T_L$, where T_L is the ground-state librational period, $T_L \simeq 2\pi/\sqrt{F_r}$, and after this time the states $k = 8$ and 9 are equally populated; for shorter switch-times many other states are populated, although states outside the island, $k \geq 10$, are significantly populated only if $T_a < T_L$.

The efficient population of other island state requires a longer switch as the coupling between relevant adiabatic states varies more rapidly in these cases; for instance 90% population if the island ground state requires a time $T_a > 18T_L$.

The final important point we emphasise is that the separatrix state can be populated from the two initial states $m \simeq \pm N_L/2$. This is important as in the original problem it means that quite different initial conditions can lead to the same outcome. In particular in Figure 1 we see that the initial states $n_0 = 56$ and $n_0 = 66$ have similar thresholds: this is because these two initial states both populate the same scarred state of the main resonant island.

4. Numerical results

In this section we compare the predictions of the theory with results obtained by numerically solving Schrödinger's equation using the method described in Leopold and Richards (1991), which includes ionisation with a decay mechanism the rates of which are obtained semi- classically. We also compare these quantal results with classical Monte Carlo simulations computed using the regularised coordinates described in Leopold and Richards (1985). For these calculations we introduce a static field in order to give a manageable basis by lowering the continuum threshold: we

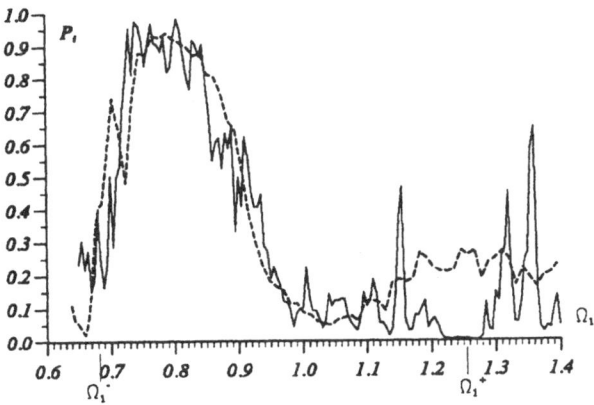

Fig. 4. Graph comparing the classical, (dashed line), and quantal, (solid line), ionisation probabilities, P_t, as a function of the scaled frequency, equation 26, computed using the Hamiltonian 22
In this example $n_0 = 40$, $F_0 = 0.04$ and $F_s = 40\text{V cm}^{-1}$, so that $\Omega_1^- = 0.680$ and $\Omega_2^- = 1.258$.

emphasis that the static field, provided it is not too large, does not qualitatively
change the dynamics in the region of the resonant island. The calculations also
include a slowly varying field envelope, $\lambda(t) = \sin(\pi t/T_m)$, $0 < t < T_m$, and are performed in the momentum gauge as this is appropriate for an atom moving through
a fixed cavity or wave guide; in all cases we choose T_m to be 100 field periods. The
Hamiltonian is, Leopold and Richards (1991),

$$H(q,p,t) = \frac{1}{2\mu}p^2 - \frac{e^2}{z} - F_s z - \frac{F\lambda(t)}{\mu\Omega}p\cos\Omega t, \quad z \geq 0, \tag{22}$$

and Schrödinger's equation is solved using a basis comprising the eigenstates of the
unperturbed Hamiltonian, $F = 0$, which includes the static field. This unperturbed
system has approximately $161(F_s/\text{Vcm}^{-1})^{-1/4}$ bound states.

If the static field is sufficiently small we may approximate its effect near the
initial state using first-order perturbation theory, so the unperturbed Hamiltonian,
equivalent to that of equation 5, and frequency are

$$H_0 = -\frac{\mu e^4}{2I^2} - \frac{3F_s I^2}{2\mu e^2}, \quad \text{and} \quad \omega(I) = \frac{\mu e^4}{I^3} - \frac{3F_s I}{\mu e^2}. \tag{23}$$

The resonant action I_r is defined by the equation $\Omega = \omega(I_r)$. In order to obtain
manageable expressions it is necessary to expand about I_r, so on putting $I = I_r + X$
and ignoring terms $O(X^3)$, we obtain the following resonant Hamiltonian

$$H_0 = \frac{\mu e^4}{I_r^2}\left\{-\frac{3}{2}\left(\frac{X}{I_r}\right)^2(1 + F_{sr}) + \frac{\alpha F_r}{1 - 3F_{sr}}\cos\phi\right\} \tag{24}$$

where $F_{sr} = F_s I_r^4 / \mu^2 e^6$ is the static field in scaled units relative to the resonant action. The number of librational states is now given by a variant of equation 8,

$$N_l \simeq \frac{8 I_r}{\pi \hbar} \sqrt{\frac{\alpha F_r}{3(1 + F_{sr})(1 - 3 F_{sr})}}.$$

In order to populate a separatrix state we showed in section 3 that it is necessary to start with an action $I_0 \simeq I_r \pm N_l \hbar / 2$. We now need to find an expression for I_0 given that we normally express the fields in units relative to the initial state. Write $I_r = x I_0$ so the equation for x is

$$\frac{1}{x} = 1 \pm \frac{4}{\pi} \sqrt{\frac{\alpha x^4 F_0}{3(1 + x^4 F_{s0})(1 - 3 x^4 F_{s0})}} = h_\pm(x). \qquad (25)$$

This equation has no simple solution, but since F_0 and F_{s0} are small the two solutions, $x_\pm$, can be found by iteration. Since the unperturbed frequency is not the same as the Kepler frequency we define the scaled frequency for this system to be

$$\Omega_1(x) = \frac{\Omega}{\omega(I_0)} = \frac{1}{x^3} \left(\frac{1 - 3 x^4 F_{s0}}{1 - 3 F_{s0}} \right), \qquad x = \frac{I_r}{I_0}. \qquad (26)$$

As an example in figure 4 we compare classical and quantal ionisation probabilities, the dashed and solid lines respectively, for $n_0 = 40$, $F_s = 40 \text{V cm}^{-1}$ ($F_{s0} = 0.0199$) and $F_0 = 0.04$ in the frequency range $0.65 \leq \Omega_1 \leq 1.4$. For these parameters $x_- = 1.1228$ and $x_+ = 0.93125$ giving $\Omega_1^- = \Omega_1(x_-) = 0.680$ and $\Omega_1^+ = 1.258$. Both classical and quantal ionisation probabilities increase sharply at Ω_1^-; for $\Omega_1 < \Omega_1^-$ most classical orbits are regular and remain below the stochastic layer at the edge of the island, although a small proportion are unstable and leak past the island through the gap at the unstable fixed point of K_0. The behaviour of the quantal system is similar as the initial state is deformed only slightly and overlaps very little with states above the island. In both cases P_i is small for $\Omega_1 < \Omega_1^-$.

At $\Omega_1 \simeq \Omega_1^-$, where $N_L \simeq 7.2$, the initial classical phase curve is adiabatically switched onto the separatrix: a high proportion of orbits are unstable and ionise. The wave function is adiabatically switched onto a separatrix state which is more strongly coupled to the ionising rotational states above the island than are the lower rotational states.

As Ω_1 increases through Ω_1^- both classical and quantal probabilities increase dramatically, then decrease as Ω_1 passes through unity. The main difference between the classical and quantal probabilities is that the latter shows sharp oscillations which we believe are produced by resonant interactions between the island and rotational states. Beyond $\Omega_1 = 1.1$ the classical probability is small but increasing: it is not clear whether the oscillations seen in the figure are real or a result of statistical errors. The quantal probability is also small, except for the sharp peaks at $\Omega_1 \simeq 1.15$ and 1.35, which we believe to be caused by the same mechanism that produces the oscillations near $\Omega_1 = 0.8$. The main point to notice, however, is that straddling the frequency $\Omega_1^+ = 1.26$ the quantal probability is almost zero, by comparison to the classical probability which is not noticeably affected by the separatrix. In this example according to equation 31 of Leopold and Richards (1991) the continuum, lowered by F_s, is classically accessible from levels $n \geq 51$, which is well above the upper edge of the resonant island.

5. Conculsions

We have provided a dynamical reason explaining why the scarred states associated with the unstable periodic orbits of resonant islands are less readily ionised than adjacent states. We do this by showing that there exists a representation in which the scarred state is less strongly coupled to the higher lying states, from which ionisation occurs, than are adjacent states. This representation exists simply because the classical motion in the vicinity of the unstable orbit is relatively slowly varying: whilst this is insufficient to prevent chaotic motion in the classical system it is sufficient to produce a local stability in the quantal system.

The theory has been developed for a periodically driven 1d hydrogen atom, but it is clear that the analysis holds for any periodically driven 1d system. The results we obtain are also in good agreement with experimental results on 3d hydrogen atoms, without a static field present, as would be expected because this system is completely degenerate so the dynamics of the 1d and 3d systems are similar, both being determined, by the same basic frequencies. Because of this similarity the theoretical and experimental positions of the scarred states are similar and the theory is able to explain why the experimental observations are relatively insensitive to the shape of the field envelope, Sauer et al (1992).

The theory presented here is limited to scars associated with the main resonance. There are resonances close to all rational scaled frequencies and scar-like structures have been observed at several of these, for example near $\Omega_0 \simeq 0.4$, 0.6 and 1.6, Koch (1992); other candidates are near $\Omega_0 = \frac{1}{2}$ and $\frac{1}{3}$, see Leeuwen et al (1985). It seems likely that a similar analysis could explain the observations at the sub-harmonic scars, $\Omega_0 \simeq r/s < 1$ if the Hamiltonian could be cast into the appropriate form; work is in progress on this extension. Theoretically there ought to be scars due to the higher order resonance, $\Omega_0 \simeq 2, 3. \cdots$, but the existence and importance of quasi-resonant states in lowering the effective density of states, Leopold and Richards (1989) suggests that these may be observable only if the initial principle quantum numbers are larger than those treated here. We are currently investigating these resonances in order to determine whether or not they can be observed in experiments on atoms with principal quantum numbers $n_0 < 100$.

Acknowledgements
It is a pleasure to thank Professor P M Koch and Dr J G Leopold for many fruitful discussions . DR thanks the SERC for a grant making this work possible.

References

Abramowitz M and Stegun I A 1964 *Handbook of Mathematical Functions* (Dover NY)
Burnett K, Reed V C, Cooper J and Knight P L 1992 Phys Rev **A45** 3347-9

Eckhardt B, Hose G and Pollak E 1989 Phys Rev **A39** 3776–93

Heller E J 1991 *Wavepacket dynamics and quantum chaology* in Les Houches lectures *Chaos and Quantum Physics* 1st–31st August 1989, Eds M -J Giannoni, A Voros and J Zinn-Justin (North Holland)

Hennenberger W C 1968 Phys Rev Lett **21** 838

Jensen R V, Sanders M M, Saraceno M and Sundaram 1989 Phys Rev Lett **63** 2771–5

Koch P M 1992 Chaos **2** 131–44

Kramers H A 1956 *Collected Scientific Papers*, (North Holland, Amsterdam)

Leeuwen van K A H, Oppen v. G, Renwick S, Bowlin J B, Koch P M, Jensen R V, Rath O, Richards D and Leopold J G 1985 Phys Rev Lett **55** 2231–4

Leopold J G and Richards D 1985 J Phys **B18** 3369–94

Leopold J G and Richards D 1989 J Phys **B22** 1931–61

Leopold J G and Richards D 1991 J Phys **B24** 1209–40

Leopold J G and Richards D 1993 To be submitted to J Phys

Percival I C and Richards D 1982 *Introduction to Dynamics* (Cambridge University Press)

Provost D and Baranger M 1993 Phys Rev Lett **71** 662–5

Reed V C and Burnett K 1990 Phys Rev **A42** 3152–5

Sauer B E, Bellermann M R W and Koch P M 1992 Phys Rev lett **68** 1633–6

Sirko L, Bellermann M R W, Haffmans A, Koch P M and Richards D 1993 Submitted to Phys Rev Lett

Waterland R L, J-M Yuan, Martens C C, Gillilan R E and Reinhardt W P 1988 Phys Rev Lett **61** 2733–6

Wintgen D and Hönig A 1989 Phys Rev Lett **63** 1467–70

Zaslavsky G M 1981 Physics Reports **80** 157–250

THE QUASICLASSICAL STATISTICAL DESCRIPTION OF QUANTUM DYNAMICAL SYSTEMS AND QUANTUM CHAOS

YU. P. VIRCHENKO
Kharkov Institute of Physics and Technology
Kharkov, 310108, Ukraine
Fax: (057) - 235 - 1738

In this lecture, I want to propose an approach which is in some sense alternative to study the chaos in quantum mechanical systems. First of all I shall explain the motivation for the construction offered below. What is the main difficulty in the study of a quantum chaos? Consider the evolution of a statistical operator $\hat{\rho}_t$ of quantum system with Hamiltonian $\hat{H}$. We have the equation of motion for $\hat{\rho}_t$,

$$i\hbar\dot{\hat{\rho}} = \left[\hat{H},\hat{\rho}\right] \ . \tag{1}$$

Then if $\hat{H}$ has a discrete spectrum we learn from the equation that all quantum mechanical observables $A_t = Sp\hat{\rho}_t\hat{A}$ are obtained by the following trigonometric sums

$$A(t) = \sum_{k,l} A_{kl} \exp\left\{\frac{i}{\hbar}(\varepsilon_k - \varepsilon_l)t\right\} \tag{2}$$

(the summation is over the spectrum ε_k). These sums determine functions which have no irreversibility in their behaviour. For example, if

$$\sum_{k,l} |A_{kl}|^2 < \infty$$

the $A(t)$'s are almost periodic functions. This follows from the famous H. Bor theorem [1] (see, for example, [2] in connection with an application to quantum mechanics). But stochasticity always is connected with irreversibility. Therefore it may seem that there is no stochasticity in quantum mechanics. From the other side there is stochasticity in classical mechanics. And if we take into account that there is the continuous transition as $\hbar \to 0$ from quantum mechanics to classical mechanics then we have a contradiction. The way out of this contradiction is that the spectrum $\{\varepsilon_k\}$ which depends on $\hbar$, is very dense as $\hbar \to 0$ and consequently the almost periodic functions (2) are very complicated. In this case we may approximate the sums (1) by integrals if the time t is not very large. The integrals describe the irreversible evolution for those values of time. Therefore the quantum stochasticity manifests itself on bounded time intervals which may be sufficiently large.

From the above arguments we have the following conclusions. The first is that we observe quantum stochasticity always in the quasiclassical limit when the density

Z. Haba et al. (eds.), Stochasticity and Quantum Chaos, 177–183.

of the spectrum is sufficiently large and the limiting classical system is stochastic at large energies. The second is that to study quantum chaos we must find the limiting spectral properties which are conserved as $\hbar \rightarrow 0$ and which make the system stochastic when it passes to the classical one. But such a method of investigation seems very complicated.

I propose to go another way. We will avoid the study of the special spectral properties of quantum mechanical systems in detail and we will attempt to investigate quantum chaos by means of classical methods. In order to do this we must formulate quantum mechanics in classical mechanical terms. First of all we notice that despite the existence of the continuous transition from quantum mechanics to classical mechanics, the statistical descriptions in quantum and in classical mechanics are essentially different. In quantum mechanics we use noncommutative probability theory on the basis of a statistical operator which acts in a Hilbert space. In classical mechanics we use commutative probability theory on the basis of a positive measure on phase space. Therefore first of all it is necessary to formulate the quantum statistical description in terms of classical probability theory, i.e. we must define the equivalent probabilistic measure determined on the phase space of the classical system. The solution of this problem is well-known in quantum optics [3]. It consists in the following. If we have the quantum mechanical system with f degrees of freedom which has a discrete spectrum, we may introduce the so-called oscillator coherent states (see, for example, [4]) $|z> = |z_1> \ldots |z_f>$, where $z = (z_1, \ldots, z_f)$,

$$z_j = (2\hbar\nu_j)^{-1/2}(q_j\nu_j + ip_j) \in \mathbb{C}$$

$$j = 1, \ldots, f ,$$

$\hspace{10cm}$ (3)

$\nu_j > 0$ are the oscillator frequencies which are arbitrary and

$$|z_j> = \exp\left\{-\frac{1}{2}|z_j|^2\right\} \sum_{n=0}^{\infty} \frac{z_j^n}{n!} |n>$$

where $|n>$ are the eigenvectors for a one-dimensional oscillator system with the energy eigenvalues $\nu_j\hbar(n+1/2)$. The states $|z>$ have the following remarkable properties:

1) The state $|z>$ is localized near the phase space point (p,q) for each $z = (z_1, \ldots, z_f)$, where $p = (p_1, \ldots, p_f)$; $q = (q_1, \ldots, q_f)$. The localization region has a size of the order of $\hbar^{1/2}$.

2) It satisfies the relations

$$q_j = <z|\hat{q}_j|z> , \qquad p_j = <z|\hat{p}_j|z>$$

$$j = 1 \ldots f$$

3) Each state $|z>$ minimizes the uncertainty relation,

$$\left[<z\left|(\Delta\hat{q}_j)^2\right|z><z\left|(\Delta\hat{p}_j)^2\right|z>\right]^{1/2} = \hbar/2$$

$$\Delta\hat{q}_j = \hat{q}_j - q_j , \qquad \Delta\hat{p}_j = \hat{p}_j - p_j ;$$

4) There is the completeness relation

$$\pi^{-J} \int_{\mathbb{C}^J} | z >< z | \, dz = 1 \qquad Sp\hat{\rho} = \pi^{-J} \int_{\mathbb{C}^J} < z|\hat{\rho}|z > dz$$

for each fixed collection of frequencies where $dz = dz_1 \ldots dz_J$, $dz_j = dp_j \, dq_j / (2\hbar)$. The arbitrary parameters ν_j characterize the relative precision for the simultaneous observation of coordinates q_j and momenta p_j in the states $|z>$.

We may introduce further the distribution density of the positive measure on the basis of coherent states [3] ,

$$F(z, \bar{z}) = \pi^{-J} < z|\hat{\rho}|z >$$

where $\hat{\rho}$ is the statistical operator of the quantum system ($\bar{z}$ is complex conjugate to z). We have $F(z, \bar{z}) > 0$, since $\hat{\rho}^+ = \rho$, $\hat{\rho} > 0$ and moreover we have

$$\int_{\mathbb{C}^J} F(z, \bar{z}) dz = 1$$

from 4) since $Sp \, \hat{\rho} = 1$.

It is remarkable that the statistical description on the basis of the density $F(z, \bar{z})$ is equivalent to the statistical description on the basis of statistical operator $\hat{\rho}$. To prove this equivalence we may consider the Glauber–Sudarshan [5] diagonal representation which is valid for any statistical operator

$$\hat{\rho} = \int_{\mathbb{C}^J} |z'> \, \rho(z') \, <z'| \, dz'$$

where $\rho(z')$ is a generalized distribution on $\mathbb{C}^J$. We obtain from this representation and from 4) the integral equation

$$F(z, \bar{z}) = \pi^{-J} \int_{\mathbb{C}^J} e^{-|z - z'|^2} \rho(z') \, dz'$$

which has a unique solution.

The density F is defined on the phase space which is the set of points (p, q) (see (3)). If we introduce the G–density, which is defined by the formula

$$G(u, \bar{u}) = (2\hbar)^J \, F\left((2\hbar)^{1/2} z, (2\hbar)^{1/2} \bar{z}\right)$$

$$u = (2\hbar)^{1/2} z$$

then we find that it tends to the probabilistic density on phase space of the corresponding classical system, if $\hbar \to 0$.

It seems at first sight that the statistical description of an arbitrary quantum system on the basis of the F (and G) – density is not convenient, since the density depends on the type of the coherent state which we select by means of setting the definite values of the parameters ν_j. Moreover, we could use non–oscillator coherent states which are more general. But this inconvenience is illusory because it is easy to see that any observable A which is defined by the formula

$$A = Sp \, \hat{\rho}\hat{A} \tag{4}$$

(here $\hat{A}$ is a self–conjugate operator) does not depend on the selection of the representation in which we calculate the trace in (4). In particular, we obtain from (4)

$$A = \pi^{-f} \int_{\mathbb{C}^f} < z|\hat{\rho}\hat{A}|z > \;. \tag{5}$$

Thus the F (and G) – density is good for the quantum statistical description. We must further find the evolution equation for the F (and G)–density and the averaging formula for an observable A in terms of the F (or G) –density. Formally these problems are sufficiently simple and we show here their solutions. But it is necessary to give a mathematical justification for the formal operations (which we do below). This justification must guarantee the convergence of the results after the calculations.

Consider the formal operator series

$$\hat{Q} = \sum_{\bar{l},l} (\bar{l}!l!)^{-1} \zeta(\bar{l},l)(a^+)^{\bar{l}} \, a^l$$

where $l = (l_1, \ldots, l_f) \in N_+^f$; $\bar{l} = (\bar{l}_1, \ldots, \bar{l}_f) \in N_+^f$; $N_+ = \{0\} \cup N, \zeta(\bar{l},l) \in \mathbb{C}$ and

$$(a^+)^{\bar{l}} = \prod_{j=1}^{f} (a_j^+)^{\bar{l}_j}$$

$$a^l = \prod_{j=1}^{f} a_j^{l_j}$$

in which a_j^+ and a_j are the creation and annihilation operators for the j-th one–dimensional oscillator

$$a_j |z> = z_j |z>, \qquad < z|a_j^+ = < z|\bar{z}_j \;.$$

For each series $\hat{Q}$ we introduce the formal functional series

$$Q(z, \bar{z}) = \sum_{\bar{l},l} (\bar{l}!l!)^{-1} \zeta(\bar{l},l) \, \bar{z}^{\bar{l}} z^l$$

where $z^l = z_1^{l_1} \ldots z_f^{l_f}$, $\bar{z}^{\bar{l}} = \bar{z}_1^{\bar{l}_1} \ldots \bar{z}_f^{\bar{l}_f}$. Such a series we call the symbol of operator $\hat{Q}$. From the other side we may construct the operator series $\hat{Q}$ for each symbol Q.

On the symbol manifold we define the operation of $*$–multiplication in the following way: Let Q and R be the symbols for some operators $\hat{Q}$ and $\hat{R}$. Then the $*$–product is calculated by the formula

$$(Q \star R)(z, \bar{z}) = \sum_{s} (s!)^{-1} (D_z^s Q)(D_{\bar{z}}^s R)$$

where $D_z^s = D_{z_1}^{s_1} \ldots D_{z_f}^{s_f}$, $s = (s_1, \ldots, s_f)$, $s_j \in N_+$ and

$$D_{z_j} = \frac{1}{2}\left(\frac{\partial}{\partial x_j} - i\,\frac{\partial}{\partial y_j}\right); \qquad D_{\bar{z}_j} = \frac{1}{2}\left(\frac{\partial}{\partial x_j} + i\,\frac{\partial}{\partial y_j}\right)$$

$$x_j = Re\ z_j, \qquad y_j = Im\ z_j\ .$$

The $*$–multiplication converts the manifold $\{Q\}$ of symbols into an associative algebra. One can prove the remarkable formula

$$< z|\hat{Q}\ \hat{R}|z > = (Q \star R)(z, \bar{z})\ ,$$

i.e. the mapping $\{\hat{Q}\} \leftrightarrow \{Q\}$ is an algebraic isomorphism. Therefore we can transform the calculations with quantum mechanical operators in the solution of the evolution equation (1) into the calculations with their symbols. From this mathematical fact we find immediately the evolution equation for the F–density

$$i\hbar \dot{F} = [H, F]_{\star} \tag{6}$$

where $H = H(z, \bar{z})$ is the symbol for the Hamiltonian $\hat{H}$ and the commutator is calculated in the algebra of symbols. In a more detailed representation the equation (6) has the form

$$i\hbar \dot{F} = -\sum_{s} (s!)^{-1} [H, F]_s \tag{7}$$

$$-[H, F]_s = (D_z^s H)(D_{\bar{z}}^s F) - (D_{\bar{z}}^s H)(D_z^s F)\ .$$

We notice that the term with $s = 0$ is absent in the sum $\sum_s$ and the terms with s for which $|s| = s_1 + \ldots + s_f = 1$ give the Poisson bracket in the $(z, \bar{z})$–variables. The right hand side of (7) we may call the "quantum Poisson–bracket". We may write the evolution equation for the G–density now in the form

$$\dot{G}_t + (i\hbar)^{-1} \sum_{s} (s!)^{-1} (2\hbar)^{|s|} \left((D_{\bar{u}}^s H)(D_u^s G) - (D_u^s H)(D_{\bar{u}}^s G)\right) = 0\ . \tag{8}$$

The last equation contains a straightforward expansion in powers of $\hbar$ and secondly we may investigate the quantum mechanical evolution from the statistical viewpoint as well as from the classical one, since the G–density has a classical limit. To do this we must construct an asymptotic expansion using the smallness of $\hbar$ for the solution of the equation (8).

To find the averaging formula we introduce the antinormally ordered operator series $\tilde{\tilde{A}}$ for the operators $\hat{A}$ in (4)

$$\hat{\tilde{A}} = \sum_{\bar{l},l} \left(\bar{l}! l! \right)^{-1} \zeta \left(\bar{l}, l \right) a^l \left(a^+ \right)^{\bar{l}}$$

$$\tilde{\zeta} \left(\bar{l}, l \right) \in \mathbb{C}$$

and the corresponding antisymbols

$$\tilde{A} = \sum_{\bar{l},l} \tilde{\zeta} \left(\bar{l}, l \right) z^l \bar{z}^{\bar{l}}.$$

Then we easily find that

$$Sp \hat{\rho} a^l \left(a^+ \right)^{\bar{l}} = \pi^{-f} \int_{\mathbb{C}^f} <z|(a^+)^{\bar{l}} \, \hat{\rho} \, a^l |z> dz$$

$$= \int_{\mathbb{C}^f} \bar{z}^{\bar{l}} \, z^l \, F(z, \bar{z}) dz$$

and consequently we have an averaging formula of the form

$$A = Sp \hat{\rho} \hat{A} = \int_{\mathbb{C}^f} F(z, \bar{z}) \, \tilde{A}(z, \bar{z}) \, dz =$$

$$= \int_{\mathbb{C}^f} G(u, \bar{u}) \, \tilde{A}(u, \bar{u}) \, du \tag{9}$$

which looks like the classical one. If we build the asymptotic expansion for small $\hbar$ each term of the corresponding expansion in the equality (9) does not depend on the coherent representation.

Thus we have obtained an alternative method for the study of the quantum dynamics and, in particular, quantum chaotic behaviour. To apply this method we must construct the asymptotic expansions uniformly approximating in time the solution of equation (10), i.e. there may not be secular terms in them. It is not a simple mathematical problem even if we know the zeroth order approximation (classical solution), but it is soluble. We write explicitly the first order approximation equation for the G–density in the (p, q)–notations

$$\dot{G}_t + \{H, G\} + \frac{\hbar}{2} \left(\nu \{D_p H, D_p G\} + \nu^{-1} \{D_q H, D_q G\} \right) = 0 \tag{10}$$

where $\{\cdot, \cdot\}$ is the ordinary classical Poisson bracket and $\nu = \nu_1 = \ldots = \nu_f$. This equation determines the main approximation for the analysis of quantum chaos.

If the approach proposed above will be sufficiently effective it may assert that there is no special problem of quantum chaos, but there is the problem of classical chaos only. However, this classical problem is not ordinary. It is necessary to study the classical evolution and, in particular, the chaotic one in the case when the G–density is smeared over a region of size $\sim \hbar^{1/2}$ and, in particular, when it is not concentrated on a definite energy surface. To understand that the problem is difficult we note that the character of classical evolution (regular or chaotic) depends on the energy.

Therefore we must mix in some cases the initial data corresponding to regular behaviour and chaotic one in an arbitrary ratio in the observation of the classical evolution.

Consider qualitatively in the framework of our approach the following simple situations:

a) the G-density is concentrated in the region with regular initial data;

b) the G-density is concentrated in the region the chaotic initial data. In case a) the evolution of the G-density is classical with the great precision, i.e. it performs the motion of the (p,q) parameters of the coherent representation along the classical trajectories. The quantum diffusion is very small. In case b) we have an intermixing of the classical trajectories. The derivatives in the equation (10) are very large after a time period coinciding with the intermixing time. The derivatives are so large that we must take into account the quantum diffusion. It is the so-called quantum suppression of classical chaotic behaviour.

In conclusion we express the hope that a consequential development of the approach proposed above will permit us to solve some new problems of quantum chaos.

References

1. N.J. Achiezer, J.M. Glazman, Theory of linear operators in Hilbert space, Nauka, Moscow, 1966 (in Russian)
2. T.Hogg and B.A. Huberman, Phys. Rev. Lett. **48**(1982)711
3. J.R. Klauder, E.C.G. Sudarshan, Fundamentals of quantum optics, W.A. Benjamin, Inc., New York, 1968
4. Peřina, Quantum statistics of linear and nonlinear optical phenomena, Kluwer Academic Publishers, 1991
5. R.J. Glauber, Phys. Rev. **131**, (1963)2766
   R.J. Glauber, Phys. Rev. Lett, **10**, 84
   E.C.G. Sudarshan, Phys. Rev. Lett. **10** (1963) 277

QUANTUM MEASUREMENT BY QUANTUM BRAIN

MARI JIBU AND KUNIO YASUE*
Research Institute for Informatics and Science
Notre Dame Seishin University
Okayama 700, Japan

Abstract. The orthodox theory of measurement due to von Neumann, London and Bauer, and Wigner is revisited from a new point of view in which physical correlates of consciousness of an observer manifest quantum coherence. The result of an observation of a quantum mechanical system in a state of superposition is stored in a state of mixture of metastable classical vacua with spontaneously broken symmetry. Goldstone bosons inherent in such long-range ordered states play the key roles to complete the act of measurement which have been long attributed to a transcendental "abstract ego."

1. Introduction

Recently we have revealed that the fundamental brain processing makes full use of quantum coherence in terms of coherent photon emission in and transmission along the hollow core of each cytoskeletal microtubule filled with water molecules (Jibu et al., 1993). Without such a nonlocal quantum coherence, the unity of consciousness cannot be realized because the physical correlates of consciousness are distributed throughout the brain (Pribram, 1966; 1971; 1991). It was difficult to explain by classical means and has been known as the "binding problem" in neurophysiological terms (Marshall, 1989; Singer, 1993; Crick and Koch, 1990). Furthermore, Penrose (1989) claimed that many brain functions are non-algorithmic and non-computational, and so beyond the conventional scope of serial and parallel neural processing. Quantum coherence can account for these functions (Penrose, 1993). This result may force us to change our practical point of view of brain processing drastically from the conventional "electrochemical" and "chaotic" regime to a new "electrophotonic" and "coherent (ordered)" one. There, behind the conventional large scale integration of dendritic networks of neurons with transmembrane ionic diffusions, the brain hides a dense microscopic subneuronal quantum optical networks with coherent photons which may be considered to take part essentially in realizing consciousness as well as unconsciousness. Although there remain many issues to be investigated further from both quantum theoretical and neurophysiological points of view, it may be of certain interest to see if such a quantum theoretical model of consciousness could provide a new insight for our understanding of the long-standing measurement problem in quantum theory. In the present paper, we revisit the orthodox theory of measurement in quantum mechanics due to von Neumann (1932), London and Bauer (1939), and Wigner (1963) from a new point of view in which physical correlates of consciousness of every observer manifest quantum

---

* Conference Speaker

Z. Haba et al. (eds.), Stochasticity and Quantum Chaos, 185–194.
© 1995 *Kluwer Academic Publishers.*

coherence.

2. Quantum Optical Coherence in Brain Processing

Microtubule is a hollow cylinder about 25 nanometers in diameter whose wall is a polymerized array of protein subunits called tubulins. Its length may range from tens of nanometers to microns, and possibly further to meters in nerve axons of large animals (Hameroff, 1987). We consider the microtubule as a hollow cylinder with radius $r_{MT} \approx 12$ and length $l_{MT} \approx 10^2$ in nanometers. The spatial region V of the hollow core of the microtubule is likely to be filled with water molecules. Let N be the total number of those water molecules and $r^j = (x^j, y^j, z^j)$ be the position of the j-th water molecule. From a physical point of view, a water molecule has average electric dipole moment $\mu = 2e_p P$ and moment of inertia $I = 2m_p d^2$ with $P \approx 0.2 \, \text{Å}$ and $d \approx 0.82 \, \text{Å}$. Here, m_p denotes the proton mass and e_p the proton charge. Due to the electric dipole moment μ, the water molecule interacts strongly with the quantized electromagnetic field in the spatial region V. Although the water molecule has molecular vibrational degrees of freedom, we are interested in only the rotational degrees of freedom, and regard it as a quantum mechanical spinning top with electric dipole moment. We restrict our discussion to the most likely case in which only the lowest and first excited energy eigenstates take part in the energy exchange between the quantized electromagnetic field. Then, one sees immediately that each water molecule can be understood as a single two-level quantum system well described by a spin variable $s^j = \frac{1}{2}\sigma$, where $\sigma = (\sigma_x, \sigma_y, \sigma_z)$ is a vector with three components given by Pauli spin matrices. Then, the Hamiltonian of the system of N water molecules is given by

$$H_{WM} = \epsilon \sum_{j=1}^{N} s_z^j \, , \tag{1}$$

where $\epsilon \approx 200cm^{-1}$ is the energy difference between the first excited state and the lowest energy state (Franks, 1972). It is convenient to describe the quantized electromagnetic field in V in terms of an electric field operator $\boldsymbol{E} = \boldsymbol{E}(\boldsymbol{r}, t) = eE(\boldsymbol{r}, t)$, where e is a constant vector of unit length pointing in the direction of linear polarization. Then, the Hamiltonian for the quantized electromagnetic field in question is given by

$$H_{EM} = \int_V E^2 d^3r \, . \tag{2}$$

Interaction between the quantized electromagnetic field and the totality of water molecules by which they can exchange energy in terms of creation and annihilation of photons is described by the total Hamiltonian

$$H = H_{EM} + H_{WM} + H_I \, , \tag{3}$$

where H_I is the interaction Hamiltonian given by

$$H_I = -\mu \sum_{j=1}^{N} \{ E^-(\boldsymbol{r}^j, t)s_-^j + s_+^j E^+(\boldsymbol{r}^j, t) \} \tag{4}$$

with

$$s_{\pm}^j = s_x^j \pm i s_y^j \qquad (5)$$

and $E^{\pm}$ are positive and negative frequency parts of the electric field operator E. Let us introduce the normal mode expansion of the electric field operator $E = E^+ + E^-$,

$$E^{\pm}(r,t) = \sum_{k} E_k^{\pm}(t) e^{\pm i(k \cdot r - \omega_k t)} , \qquad (6)$$

and collective dynamical variables $S_k^{\pm}(t)$ and S for water molecules,

$$S_k^{\pm}(t) \equiv \sum_{j=1}^{N} s_{\pm}^j(t) e^{\pm i(k \cdot r^j - \omega_k t)} , \qquad (7)$$

$$S \equiv \sum_{j=1}^{N} s_z^j , \qquad (8)$$

where ω_k denotes the proper angular frequency of the normal mode with wave vector k. Then, the total Hamiltonian (3) becomes

$$H = H_{EM} + \epsilon S - \mu \sum_{k} (E_k^- S_k^- + S_k^+ E_k^+) . \qquad (9)$$

A physical inspection of the form of the total Hamiltonian (9) reveals that it manifests a dynamical symmetry property not evident in the ground state and so the resulting quantum dynamics is known to involve certain long-range order creating phenomena due to spontaneous symmetry breaking (Ricciardi and Umezawa, 1967; Stuart et al., 1978; 1979; Jibu and Yasue, 1992A; 1992B; 1993A; 1993B). The spatial dimension of this long-range order, that is, the coherence length l_c can be estimated to be inversely proportional to the energy difference ϵ, obtaining $l_c \approx$ hundreds of microns. Among those long-range order creating phenomena we may find a specific one in which the collective dynamics of the majority of water molecules inside the microtubule cylinder V can give rise to cooperative spontaneous emission of photons without any pumping light. Any incoherent and disordered energy distribution among the water molecules due to the macroscopic thermal dynamics of the polymerized array of protein subunits (i.e., tubulins) forming the wall of microtubule cylinder can be gathered collectively into coherent and ordered dynamics ready to emit coherent photons cooperatively. This laser-like process of coherent photon emission without pumping light was first introduced by Dicke (1954) and called photon echo or superradiance. Let us investigate the photon echo in the microtubule cylinder V starting from the total Hamiltonian (9). We assume for simplicity that only one normal mode with a specific wave vector, say k_0, has a proper angular frequency ω_{k_0} resonating to the energy difference ϵ between the lowest and first excited energy eigenstates. Namely we have

$$\epsilon = \hbar \omega_{k_0} , \qquad (10)$$

and all the other normal modes are neglected. In the conventional laser theory, this is known as a single mode laser. Since we have only one normal mode with wave vector k_0, we may omit all the wave vector indices of the dynamical variables. Then, the total Hamiltonian (9) becomes

$$H = H_{EM} + \epsilon S - \mu(E^- S^- + S^+ E^+) \ . \tag{11}$$

The corresponding Heisenberg equations of motion for the three collective dynamical variables, S and $S^\pm$, for water molecules and the two variables, $E^\pm$, for the quantized electromagnetic field are given by:

$$\frac{dS}{dt} = -i\frac{\mu}{\hbar}(E^- S^- - S^+ E^+) \ , \tag{12}$$

$$\frac{dS^\pm}{dt} = \pm i\frac{2\mu}{\hbar}SE^\mp \pm i\frac{\epsilon}{\hbar}S^\pm \ , \tag{13}$$

and

$$\frac{dE^\pm}{dt} = \pm i\frac{2\pi\epsilon\mu}{\hbar V}S^\mp \ . \tag{14}$$

Because of the short length of the microtubule cylinder $l_{MT} \approx 10^2$-10^3 in nanometers, the pulse mode propagating along the hollow core of the microtubule cylinder stays in the cavity region V only for a short transit time $t_{MT} \equiv \frac{l_{MT}}{c}$, where c stands for the speed of light. As this transit time of the pulse mode is much shorter than the characteristic time of thermal interaction due to the disordered environment, the system of water molecules and quantized electromagnetic field in this single mode photon echo is free from thermal loss and can be considered as a closed system well described by the Heisenberg equations of motion (12)-(14). It is known that a semi-classical approximation of the quantum dynamical system of photon echo yields approximative solutions to the Heisenberg equations (12) to (14) (Agarwal, 1971). The intensity of coherent photon emission in the microtubule cylinder due to photon echo can be given in this approximation by

$$I = \frac{\hbar^2}{(4t_R\mu)^2}\, sech^2\frac{t - t_0}{2t_R} \ , \tag{15}$$

where $t_R = \frac{c\hbar^2 V}{4\pi\mu^2\epsilon N l_{MT}}$ and $t_0 = t_R \ln 2N$ denotes the life time and delay time of the photon echo, respectively. Notice that the intensity of photon echo is proportional to N^2 and its delay time is inversely proportional to N. These facts are characteristic to the long-range order creating process involving N water molecules. We have shown the possibility of a completely new mechanism of fundamental brain functioning in terms of coherent photon emission by photon echo in microtubules. Unlike a laser, photon echo is a specific quantum mechanical ordering process with a characteristic time much shorter than that of thermal interaction. Therefore, microtubules may be thought of as ideal optical encoders providing a physical interface between 1) the conventional macroscopic system of classical, disordered and incoherent neural dynamics in terms of transmembrane ionic diffusions as well as thermally perturbed molecular vibrations and 2) the yet unknown microscopic photonic computing network system of ordered and coherent quantum dynamics free from thermal noise and loss. However, it is not clear whether the pulse mode coherent photons created in the microtubule cylinder by photon echo can be safely transmitted, preserving its long-range coherence. It seems most likely that even coherent photons emitted by photon echo will lose immediately their coherence and long-range order due to the noisy thermal environment. In any case, we absolutely require some quantum dynamical mechanism to maintain the coherent transmission of photons in the yet unknown microscopic photonic computing network in brain cells (Hameroff, 1974). Let us suppose that the pulse mode coherent photons are created in a small segment of the microtubule cylinder by photon echo. Then, those photons propagate along the longitudinal axis of the microtubule cylinder which is supposed to coincide with the z- axis. If the region V inside the microtubule cylinder were maintained at vacuum, they would transmit through the region just as they do along a waveguide. However, the region is filled up with water molecules, and it is not evident at all whether the pulse mode coherent photons can be safely transmitted through the region without absorption or loss of coherence. Let us consider the Maxwell equation for the scalar electric field $E = E(z,t)$, standing for the pulse mode coherent photons propagating along the z-axis coupled to the collective dynamical variables $S^{\pm}$ of water molecules inside the microtubule cylinder,

$$\frac{\partial E^{\pm}}{\partial z} + \frac{\partial E^{\pm}}{\partial t} = \mp i \frac{2\pi\epsilon\mu}{\hbar V} S^{\mp} . \tag{16}$$

This equation is valid under the condition that the collective dynamical variables $S^{\pm}$ are slowly varying, that is,

$$\frac{\partial S^{\pm}}{\partial t} \ll i\omega S^{\pm} . \tag{17}$$

Then, taking the expectation of those quantum mechanical variables in this Maxwell equation and the Heisenberg equations of motion (12) and (13), eliminating all the expectation values of those variables referring to water molecules, and introducing new variables for the scalar electric field by

$$\theta^{\pm}(z,t) = \frac{2\mu}{\hbar} \int_{-\infty}^{t} E^{\pm}(z,u)du , \tag{18}$$

we can obtain dynamical equations for the electromagnetic field in the region V inside the microtubule cylinder

$$\frac{\partial^2 \theta^{\pm}}{\partial \tau \partial \zeta} = -\sin\theta^{\pm} . \tag{19}$$

Here, $\tau = \sqrt{\frac{2\pi\epsilon\mu^2 N}{\hbar^2 V}}(t - \frac{z}{c})$ and $\zeta = \sqrt{\frac{2\pi\epsilon\mu^2 N}{\hbar^2 V}}\frac{z}{c}$. This is a typical nonlinear partial differential equation called the Sine-Gordon equation, and its exact solution gives rise to an explicit form of the time evolution of the scalar electric field E in the region V inside the microtubule cylinder

$$E = \sqrt{\frac{2\pi\epsilon N v_0}{V(c - v_0)}} sech\sqrt{\frac{2\pi\epsilon\mu^2 N v_0}{\hbar^2 V(c - v_0)}}(t - \frac{z}{v_0}) . \tag{20}$$

This is nothing but a soliton solution and tells us that the pulse mode photons propagate along the dielectric waveguide of the microtubule cylinder filled up with water molecules with a certain constant speed v_0 less than the speed of light in vacuum c. It is important to see that the pulse form of the soliton solution is kept unchanged due to the nonlinearity of the Sine-Gordon equation. It was found that microtubules play the role of dielectric waveguides and that pulse mode coherent photons propagate through them as if they were perfectly transparent. This phenomenon is termed self-induced transparency and known to be a typical non-linear effect in quantum optics (McCall and Hahn, 1967). The microtubule may be an ideal microscopic optical device for use as a perfectly transparent pathway for pulse mode photons, free from thermal noise and loss. Combined with photon echo, this self-induced transparency of the microtubule allows us to conclude that the brain may be essentially a dense assembly of microscopic and elaborated optical computing networks of microtubules in the cytoskeletal structure of brain cells. Coherent photon emission and transfer in each microtubule are ensured by photon echo and self-induced transparency characteristic to long-range ordering phenomena in quantum dynamics.

3. Quantum Brain Dynamics and Quantum Measurement

The conventional point of view of the mechanism of fundamental brain processing has been based on the neural network model. There, a large scale integration of dendritic networks of neurons with transmembrane ionic diffusions has been believed to be a sole basis for the brain processing. We coined an alternative point of view based on the subneuronal quantum optical network model. As we have seen, a dense assembly of cytoskeletal network of microtubules provides a huge subneuronal signal processing mechanism in terms of coherent photon emission and transfer in microtubules. It may open a new possible way to understand what consciousness and unconsciousness are. Let us make a hypothesis here that consciousness is realized by the quantum dynamics of the system of water molecules and quantized electromagnetic field within and among cytoskeletal network of microtubules. We call it "quantum brain dynamics" and abbreviate to "QBD." The basic idea of QBD was first given by Ricciardi and Umezawa (1967), and then have been developed by Stuart et al. (1978; 1979) and by Jibu and Yasue (1992A; 1992B; 1993A; 1993B). We sketch here the fundamental brain processing within the realm of QBD. The conventional neural network dynamics of macroscopic nature also takes part therein, and we call it "classical brain dynamics" and abbreviate to "CBD." Thus, the brain turns out to be a mixed physical system with QBD system and CBD system as it has been emphasized by Stuart et al. (1979). Let us consider in what way consciousness becomes conscious of some external event. Among various external events those which do not cause any change in QBD system are certainly out of consciousness. In other words, consciousness can be conscious of a certain external event only when the physical state of QBD system evolves suffering from some physical action caused by the external event. For us human beings, most of external events come into consciousness through visual and auditory sensations. It is well known in both cases that the external event puts energy of macroscopic level into sensory organs, and induces macroscopic physical phenomena such as the transmembrane ionic diffusion in CBD system. Dynamics of neuronal dendritic networks or neural networks suffers from the macroscopic energy (of light or sound) by receiving the pulse mode membrane electric potential generated by sensory organs. Then, various functional proteins embedded in each cell membrane convert energy of the electric potential into that of molecular conformation in the cytoskeletal structure. These energies are both of incoherent nature which might be dissipated to thermal energy. However, as we have seen in the preceding section, such an incoherent energy of molecular conformation in the cytoskeletal structure resulting from the external event is converted in part into coherent quantum dynamics of the system of water molecules and electromagnetic field within and among microtubules by means of photon echo. Namely, coherent photons are created by incoherent and disordered energy of molecular conformation arising indirectly from the physical action of the external event, and then take part in QBD system. The state of QBD system becomes changed, therefore, since these coherent photons affects directly QBD. This might be a major pathway along which consciousness becomes conscious of some external event. Next, we will consider a typical physical process of measurement form the point of view of von Neumann (1932), London and Bauer (1939), and Wigner (1963). A good account of it is given by Nelson (1967). There, unlike the orthodox Copenhagen interpretation, the physical process of measurement is fully considered as a quantum mechanical interaction process between two systems, that is, the system to be measured or observed and the system to measure or observe the former. They are called the system and apparatus, respectively, according to the custom. Since both of them are made of atoms and molecules, they are in principle quantum mechanical systems. The physical process of measurement or observation is nothing but a quantum mechanical process of interaction between the system and apparatus. Suppose the system is in a state ψ of superposition of the two orthogonal states ψ_1 and ψ_2, so that $\psi = \alpha_1 \psi_1 + \alpha_2 \psi_2$. When we observe the system to determine whether it is in the state ψ_1 or the state ψ_2 by measuring the corresponding observable, we know that the probabilities are respectively $|\alpha_1|^2$ and $|\alpha_2|^2$ thanks to the Copenhagen interpretation of quantum mechanics. But the act of observation or measurement is certainly to know which state the system is in. The expectated value of any observable of the system, say A, is given by a sesquilinear form

$$
\begin{aligned}
(\psi, A\psi) &= (\alpha_1 \psi_1 + \alpha_2 \psi_2, A(\alpha_1 \psi_1 + \alpha_2 \psi_2)) \\
&= |\alpha_1|^2 (\psi_1, A\psi_1) + |\alpha_2|^2 (\psi_2, A\psi_2) + \overline{\alpha}_1 \alpha_2 (\psi_1, A\psi_2) + \overline{\alpha}_2 \alpha_1 (\psi_2, A\psi_1) ,
\end{aligned}
\tag{21}
$$

where (,) denotes the Hilbert space inner product. The apparatus is a specific physical system designed to measure whether the system is in the state ψ_1 or ψ_2 by coupling to the system, and that after the interaction the system plus apparatus is in the state

$$
\varphi = \alpha_1 \psi_1 \otimes \chi_1 + \alpha_2 \psi_2 \otimes \chi_2 ,
\tag{22}
$$

where χ_1 and χ_2 are orthogonal states of the apparatus. This represents the state of the system plus apparatus after the interaction, and the expected value of any observable A of the system becomes

$$
(\varphi, A \otimes I \varphi) = |\alpha_1|^2 (\psi_1, A\psi_1) + |\alpha_2|^2 (\psi_2, A\psi_2) ,
\tag{23}
$$

where I denotes the identity operator. The interaction with the apparatus thus changes the system to behave like a mixture of ψ_1 and ψ_2 rather than a superposition. However, this fact is no more than the Copenhagen interpretation, and letting the system interact with any apparatus can never give us more information. In what way, indeed, could we know whether the system is in the state ψ_1 or ψ_2? Of course, if we knew that after the interaction the apparatus is in the state χ_1, we would know that the system is in the state ψ_1. "But how do we tell whether the apparatus is in the state χ_1 or χ_2? We might couple it to another apparatus, but this threatens an infinite regress." (Nelson, 1967) It is von Neumann's point of view that such an infinite regression of the "boundary" between the system and apparatus can be continued deep into the observer's sensory organs, and even to his brain. Therefore, von Neumann needed the existence of mind or abstract ego so that the measurement process would be closed. After the abstract ego has become aware of the state χ_1 or χ_2, the act of measurement is complete. In practice, the role of abstract ego is played by the consciousness of an observer. Then, naive questions may arise:

1. How does the consciousness of the actual observer become conscious of the state χ_1 or χ_2?

2. What is the ultimate apparatus under direct observation by the consciousness to which those states χ_1 and χ_2 belong?

Those questions have been left open simply because the physical correlates and mechanism of consciousness were not known. However, after developing a new quantum theoretical framework of fundamental brain processing, we may be in a better position to put those questions into serious consideration.

4. Quantum Measurement by Quantum Brain

Let us consider the total Hamiltonian (3) of QBD which is invariant under the continuous rotation group around the z-axis. We are now interested in the stable ordered state and the general quantum mechanical structure of QBD system. It is a direct consequence of the symmetry property of the total Hamiltonian (3) that the vacuum state is infinitely degenerate. Namely, just like the non-Abelian gauge theory, we have infinitely many classical vacua (Yasue, 1978). Each classical vacuum state may be characterized by commutative dynamical variables which are nothing but time-independent solutions to the Heisenberg equation (Stuart et al., 1979). Especially, the spin variables s^j's manifest constant uniform configuration

$$s^j = b, \qquad (24)$$

where b is a constant numerical vector. Those classical vacua are metastable ordered states which violate the original rotational symmetry of QBD system, and the effect of the spontaneous symmetry breaking must be taken into account (Ricciardi and Umezawa, 1967; Stuart et al., 1978; 1979). Namely, it is not sufficient to consider only a (Fock) Hilbert space of all the excited states constructed on a single vacuum state of spontaneous symmetry breaking type characterized by a classical dynamical variable. We have to introduce a larger Hilbert space indexed by the two-dimensional rotation group S_2 around the z-axis. The proper Hilbert space for QBD system with the total Hamiltonian (3) invariant under S_2 is given by a continuous direct sum

$$\mathcal{H} = \int_{S_2}^{\oplus} \mathcal{H}_\theta d\mu(\theta) \, , \tag{25}$$

where $\mathcal{H}_\theta$ denotes the Hilbert space of all the excited states on a single vacuum state characterized by each value θ of rotation in S_2 and μ is the invariant measure of the rotation group S_2. The Hilbert space $\mathcal{H}$ decomposes as a continuous direct sum of Hilbert spaces $\mathcal{H}_\theta$'s distinguished exactly by the parameter θ in such a way that two states from different subspaces do not have a superposition and those from the same one do (Araki, 1980). In this sense, the Hilbert spaces $\mathcal{H}_\theta$'s are said to be coherent subspaces. Linear combinations of vectors from different coherent subspaces give rise only to mixtures in which only one state remains after knowing which state the QBD system is actually in. This is simply a purely probabilistic process of exclusive events. We consider QBD system itself as the ultimate apparatus under direct observation by the consciousness. Then, two orthogonal states χ_1 and χ_2 belong to the Hilbert space $\mathcal{H}$, and so the linear combination (22) becomes automatically a mixture. This is a possible answer to the second naive question. The consciousness was nothing but dynamics of QBD system in our point of view. Then, how does it become conscious of the state χ_1 or χ_2? It is crucial that a memory is stored in QBD as a metastable vacuum state characterized by a classical dynamical variable (Stuart et al, 1978; 1979). Therefore, two states χ_1 and χ_2 are necessarily such metastable states with long-range order. Otherwise, the result of measurement will be lost almost instantaneously. If the state happens to be χ_1 while in the mixture (22), the result of this observation is stored in this metastable vacuum state χ_1 itself. Of course, there are plenty of metastable states corresponding to memories of other events already stored before the act of this measurement. Dynamics of QBD system (and so the consciousness) is then represented by creation and annihilation of energy quanta on those vacuum states. Then, the consciousness of the observer becomes conscious of the result of this observation χ_1, if the presence of the metastable vacuum state χ_1 can affect such a dynamics of QBD system. This is not so difficult thanks to the Nambu-Goldstone theorem in quantum field theory (Umezawa, 1993): As the metastable vacuum state χ_1 is of spontaneous symmetry breaking type, long-range coherent waves with zero energy requirement, that is, Goldstone bosons are created by any small amount of disturbance. These Goldstone bosons are inherent in the metastable vacuum state (and so in the result of this observation), and take part in dynamics of QBD system (and so in consciousness of the observer). It is in this sense the observer can know the result of the observation of the system and the act of measurement is complete. This might be a possible answer to the first naive question.

Acknowledgements The authors wish to thank Professors H. Umezawa, Y. Takahashi, L. M. Ricciardi and C. I. J. M. Stuart for providing us with a quantum theoretical framework for living matter. They also wish to thank Professors K. H. Pribram, S. Hameroff and S. Hagan for active collaboration. An illuminating discussion with Professor T. Nakagomi helped us much.

References

1. Agarwal, G. S., 1971, Master-equation approach to spontaneous emission III: many-body aspects of emission from two-level atoms and the effect of inhomogeneous broadening. Phys. Rev. $A4$, 1791-1801.
2. Araki, H., 1987, On superselection rules, in: *Proc. 2nd Int. Symp. Foundations of Quantum Mechanics*, Y. Ohnuki et al. (eds.) (Physical Society of Japan, Tokyo).
3. Crick , F. and Koch, C., 1990, Towards a neurobiological theory of consciousness. Seminars in the Neurosciences *2*, 263-275.
4. Dicke, R. H., 1954, Coherence in spontaneous radiation processes. Phys. Rev. *93*, 99-110.
5. Franks, F., 1972, *Water: A Comprehensive Treatise* (Plenum, New York).
6. Hameroff, S. R., 1974, Chi: a neural hologram? Am. J. Chi. Med. *2*, 163-170.
7. Hameroff, S. R., 1987, *Ultimate Computing: Biomolecular Consciousness and Nano Technology* (North-Holland, Amsterdam).
8. Jibu, M., Hagan, S., Hameroff, S. R., Pribram, K. H. and Yasue, K., 1993, Quantum optical coherence in cytoskeletal microtubules: implications for brain function, in press.
9. Jibu, M. and Yasue, K., 1992A, A physical picture of Umezawa's quantum brain dynamics, in: *Cybernetics and Systems Research '92*, R. Trappl (ed.) (World Scientific, Singapore).
10. Jibu, M. and Yasue, K., 1992B, The basics of quantum brain dynamics, in: *Proceedings of the First Appalachian Conference on Behavioral Neurodynamics*, K. H. Pribram (ed.) (Center for Brain Research and Informational Sciences, Radford University, Radford, September 17- 20).
11. Jibu, M. and Yasue, K., 1993A, Intracellular quantum signal transfer in Umezawa's quantum brain dynamics. Cybernetics and Systems: An International Journal *24*, 1-7.
12. Jibu, M. and Yasue, K., 1993B, Introduction to quantum brain dynamics, in: *Nature, Cognition and System III*, E. Carvallo (ed.) (Kluwer Academic, London).
13. London, F. and Bauer, E., 1939, *Théorie de l'Observation en Mécanique Quantique* (Hermann, Paris).
14. McCall, S. L. and Hahn, E. L., 1967, Self-induced transparency by pulsed coherent light. Phys. Rev. Lett. *18*, 908-911.
15. Marshall, I. N., 1989, Consciousness and Bose-Einstein condensates. New Ideas in Psychology *7*, 73-83.
16. Nelson, E., 1967, *Dynamical Theories of Brownian Motion* (Princeton University Press, New Jersey).
17. Penrose, R., 1989, *The Emperor's New Mind* (Oxford University Press, London).
18. Penrose, R., 1993, *Shadows of the Mind* (Oxford University Press, London).
19. Pribram, K. H., 1966, Some dimensions of remembering: steps toward a neuropsychological model of memory, in: *Macromolecules and Behavior*, J. Gaito (ed.) (Academic Press, New York).
20. Pribram, K. H., 1971, *Languages of the Brain* (Englewood Cliffs, New Jersey).
21. Pribram, K. H., 1991, *Brain and Perception* (Lawrence Erlbaum, New Jersey).
22. Ricciardi, L. M. and Umezawa, H., 1967, Brain and physics of many-body problems. Kybernetik *4*, 44-48.
23. Singer, W., 1993, Synchronization of cortical activity and its putative role in information processing and learning. Ann. Rev. Physiol. *55*, 349-374.
24. Stuart, C. I. J. M., Takahashi, Y. and Umezawa, H., 1978, On the stability and non-local properties of memory. J. Theor. Biol. *71*, 605-618.
25. Stuart, C. I. J. M., Takahashi, Y. and Umezawa, H., 1979, Mixed-system brain dynamics: neural memory as a macroscopic ordered state. Found. Phys. *9*, 301-327.
26. von Neumann, J., 1932, *Mathematishe Grundlagen der Quantenmechanik* (Springer, Berlin).
27. Wigner, E. P., 1963, The problem of measurement. Am. J. Phys. *31*, 6-15.
28. Yasue, K., 1978, Quantum decay process of meta-stable vacuum states in SU(2) Yang-Mills theory: a probability theoretical point of view. Phys. Rev. *D18*, 532-541.

TIME REVERSAL AND GAUSSIAN MEASURES IN QUANTUM PHYSICS

J.C. ZAMBRINI
CFMC Av. Prof. Gama Pinto 2,
1699 Lisbon Codex, Portugal

Content

1. Introduction

Nowadays, an important portion of the physicist's community believes that Quantum Mechanics should contain all of Classical Mechanics, including chaotic phenomena.

Formulated in this vague way, there is little evidence, as yet, to substantiate such a conviction. The recent orientations of research regarding chaos and quantum physics have been progressively restricted to the semiclassical regime where it is expected that well posed problems can be solved. But even in this more restricted area, puzzling observations have been made. If one tries to quantize a classical chaotic system, namely one which exhibit, at least, a remarkable sensitivity of its final state which respect to the initial one, one gets immediately into trouble. Besides, as was observed long ago by Einstein, it could very well be that the usual rules of quantization are not valid anymore in this situation.

Two lines of thought will guide the present essay, considerably more restricted in its ambition than the quantization of chaotic systems. Actually only elementary integrable systems will be considered here, for quantization. Nevertheless the strategy proposed, although very old in its principle (1931 !) is not at all the usual one. Our first line of thought is the role of probability in classical chaotic systems. Although the initial study of such systems has often been done from a resolutely deterministic point of view, the technical use of probability increased continuously during the last 10 years. There should be no surprise here: it was proved, already some time ago [1] that there is no way to distinguish between a deterministic but chaotic orbit and the sample function $z \cdot (\omega)$ of some random process, where ω denotes a fixed value of the chance parameter in a set Ω, typically a collection of paths $\Omega = \left\{ \omega : [0 \ T] \to \mathbf{R}^d \right\}$ with some regularity properties.

Z. Haba et al. (eds.), Stochasticity and Quantum Chaos, 195–207.
© *1995 Kluwer Academic Publishers.*

This classical probabilistic aspect has nothing to do, however, with the way probability enters into quantum physics. As it is well known, Feynman's reinterpretation of the Born interpretation for the wave function ψ of a system involves a "probabilistic" representation of ψ, an integral over a space of paths or "path integral" [2]. The underlying paths are non differentiable as long as the Planck constant $\hbar$ is different from zero. In the classical limit, however, there are those regular paths solving the Euler-Lagrange equations of the system. One may hope that Feynman approach could be generalized up to the point to include the extra randomness, of purely classical nature, that we have to face when trying to quantize chaotic dynamical system. Then, it would be possible to study the competition of the quantum and classical fluctuations. There is still, however, a long way to go.

Our second line of thought is the role of time reversal in quantum physics. An interesting numerical observation is that, even for systems classically chaotic, the result of quantization seems to be time reversible, although the classical system is obviously not. This suggests that the time reversibility of quantum mechanics is deeper than the nature of the underlying classical motions and that, if some kind of extension of quantum theory is needed to approach chaos (there are some evidences in this direction today), the time symmetric nature of this extended theory should be carefully preserved.

It is worth observing that the problems associated with time reversal are already quite subtle in classical, nonchaotic, physics. A famous one, which should be kept in mind for what follows is the problem of interpretation of the dominance of retarded electromagnetic radiation over advanced radiation in the universe. It is fair to say that the origin of this "arrow of time" is still unsolved.

The traditional treatment of relativistically invariant wave equations involves both retarded (or positive mass energy) and advanced (or negative mass energy) solutions, corresponding to the two possible directions of time. One selects, in general, special ("causal") boundary conditions in order to eliminate the "unphysical" advanced solutions. Then, certainly, an arrow of time manifests itself, but in a way physically unconvincing, except if one could prove that the chance to pick another kind of boundary conditions is quasi nil.

A more interesting approach is the famous "Action-at-a-distance theory" of Wheeler and Feynman [2] which preserves the time-symmetry by starting from a linear combination of advanced and retarded solutions to the relevant wave equation and then makes the agreement with time asymmetric observations through the use of any (not necessarily causal) boundary conditions.

We are going to show an analogy between this way to deal with time reversal and the simplest possible quantization procedure. It will be possible because we adopt a (semiclassical, here) method of quantization inspired by E. Schrödinger in 1931 and developed since the mid eighties under the somewhat optimistic name of "Euclidean Quantum Mechanics" [3 and 6].

The method is Euclidean therefore the problem of quantization amounts to the construction of (Gaussian, here) probability measures on path spaces, associated with the quantization of some classical systems. This should be just regarded as a rigorous version of Feynman's method of quantization. Of course this rigour has a price: we are only dealing here with an analogy of quantum physics, but quite a strong one.

2. Quadratic Lagrangians and Gaussian measures

Let us start from the data of an Euclidean Lagrangian, for example

$$\overline{L}(\omega,\dot{\omega}) = \frac{1}{2}\,|\dot{\omega}|^2 + V(\omega) \tag{2.1}$$

where the usual change of sign of the kinetic energy term has been hidden with a multiplication by (-1) of the whole Lagrangian. V is any bounded from below scalar potential. We will work in one dimension, only for notational simplicity, and will always denote by a bar Euclidean quantities.

Let us consider a solution of the heat equation on the time interval $[\bar{},0]$

$$\begin{cases} \dot{\eta} = H\eta \\ \eta_\chi(0) = \chi \end{cases} \tag{2.2}$$

where H is the Hamiltonian operator associated with the Lagrangian (2.1), namely

$$H = -\frac{\hbar^2}{2}\Delta + V\;. \tag{2.3}$$

Notice that, as a Cauchy value problem, Eq (2.2) would not seem to be very natural choice, because of the absence of the minus sign in the left hand side. This is why we have chosen a final boundary condition for this P.D.E., to be solved for $t < 0$, so that the problem remains well posed.

The Feynman-Kac formula is the following path integral representation of the (regular) solutions of the PDE (2.2)

$$\eta_\chi(y,t) = \int_{\Omega y} \chi(\omega(0))e^{-\frac{1}{\hbar}\int_t^0 V(\omega(\tau))d\tau}\,d\mu^W(\omega) \tag{2.4}$$

where $d\mu^W(\cdot)$ denotes an integration with respect to the Wiener measure, namely the well-defined version of the formal exponential of the kinetic energy term in Feynman's action functional [4]. In this sense, the Lagrangian $\overline{L}$ underlying Eq (2.4) is indeed the one given in (2.1). The space Ωy on which the integral (2.4) is computed is the space of all continuous paths starting from the configuration y at the (past) time t,

$$\Omega y = \{\omega : C([t,0] \to \boldsymbol{R})\ \text{ such that }\ \omega(t) = y\}.$$

The probabilistic meaning of the past condition $\omega(t) = y$ is that the right hand side of Eq (2.4) is, actually, a conditional expectation, taking into account in the calculation of the mean only those paths starting from y.

As is well known, in the semiclassical regime, the relevant action functional is not the one involved in the path integral (2.4) but the quadratic one associated with its second variation,

$$\delta^2 \overline{S}[q(.)] = \int_t^u \left\{ \frac{1}{2} \dot{q}^2(\tau) + \frac{1}{2} \nabla^2 V(z^0(\tau)) q^2(\tau) \right\} d\tau \tag{2.5}$$

whose integrand defines the semiclassical Lagrangian $\overline{L}_{sc}(\dot{q}, q, \tau)$. As for the path integral (2.4), although the action (2.5) is, a priori, well defined only for trajectories of class C^1, it is used formally in a Feynman-Kac formula to define the semiclassical measures.

In (2.5), $z^0(\tau)$ denotes a classical minimizing trajectory for the first variation. If we assume that the potential V is such that $\nabla^2 V(z^0(\tau))$ is positive, the right–hand side of (2.5) can be written as the norm $(q, q)_1$ of the Sobolev space $H_1([t, u])$ such that

$$(\phi, \varphi)_1 = \int_t^u \phi(\tau)(C^{-1}\varphi)(\tau) d\tau \tag{2.6}$$

where C denotes the covariance operator associated with $\delta^2 \overline{S}[q(.)]$, namely

$$C = \left(-\frac{d^2}{d\tau^2} + \nabla^2 V(z^0(\tau)) \right)^{-1}.$$

Implicitly, in Eq (2.6), some boundary conditions have been chosen for C^{-1} (due to the integration by parts), for example the domain of C^{-1} was $\mathcal{D}^0(C^{-1}) = C_0^\infty([t, u])$. But this is not the only choice or even an especially natural one (regular textbooks on Functional Integration are rather elusive on this question. Cf. [5] for example). A more natural domain will be: $\mathcal{D}(C^{-1}) = \{q(.)$ regular on [t,u], such that $q(t) = y$ and $F_u.\dot{q}(u) - \dot{F}_u q(u) = \delta$ for any coefficients $F_u, \dot{F}_u$ with $|F_u| + |\dot{F}_u| > 0 \}$ and δ is an unspecified extra coefficient, whose meaning will be clarified later.

Then the second variation functional becomes

$$\delta^2 \overline{S}[q(.)] = \int_t^u \overline{L}_{sc}(\dot{q}, q, \tau) d\tau - \frac{1}{2} \dot{F}_u F_u^{-1} q^2(u) - \delta F_u^{-1} q(u) \tag{2.7}$$

For every pair of coefficients $F_u, \dot{F}_u$, we obtain in this way a distinct self-adjoint extension of C^{-1} equipped with $\mathcal{D}^0(C^{-1})$.

The gradient of the family of future boundary conditions involved in (2.7) provides a final momentum field:

$$\dot{q}(u)_{|q(u)=z} = \dot{F}_u F_u^{-1} z + F_u^{-1} \delta \equiv B_u(z) \tag{2.8}$$

It can be regarded as the semiclassical interpretation of the initially given action functional. Before showing in which sense, let us notice that the graph of B_u, i.e. the collection of points, in the classical phase space, of the form $(z, B_u(z))$ constitutes a (final) Lagrangian manifold.

As well known in the context of Feynman's path integrals, more singular Lagrangian manifolds will be needed later, notably $q(u) = z$.

Now let us interpret the pair of coefficients $\begin{pmatrix} F_u \\ \dot{F}_u \end{pmatrix}$ as the final boundary conditions of a certain solution of the homogeneous Jacobi equation associated with $\delta^2 \overline{S}$, i. e. for $\tau \in [t, u]$,

$$-\frac{d^2}{d\tau^2} q(\tau) + \nabla^2 V(z^0(\tau)) q(\tau) = 0 \tag{2.9}$$

whose set of solutions is, here, a two dimensional vector space. Let us denote by $F(\tau)$ the solution of Eq (2.9) for the above mentioned pair of (final) boundary conditions.

It is well known that, for boundary conditions like the ones of $\mathcal{D}(C^{-1})$ (which are of the Sturm-Liouville "unmixed" kind), the integral kernel of the covariance operator C is of the symmetric form

$$K_t(\tau_1, \tau_2) = \begin{cases} F(\tau_2) F_t^*(\tau_1), & t < \tau_1 < \tau_2 < u \\ \\ F(\tau_1) F_t^*(\tau_2), & t < \tau_2 < \tau_1 < u \end{cases} \tag{2.10}$$

where $F_t^*(.)$ is another solution of the same Jacobi equation, independent of the first one, and chosen such that it satisfies the homogeneous part of the left boundary condition needed for $\mathcal{D}(C^{-1})$, namely $F_t^*(t) = 0$. One can also pick a Wronskian $W \equiv W(F, F_t^*)$ such that $\dot{F}_t^*(t) = 1$. The condition $F_t^*(t) = 0$ expresses physically the fact that the underlying Gaussian processes are conditioned to start from a point at initial time t, like the processes involved in the path integral of $\eta_\chi(y, t)$. The future (or right hand) boundary condition defines a manifold necessary for our semiclassical purpose, $q_{t,y}(\tau) \in \mathcal{D}(C^{-1})$. One proves the following [6]: There is a family of Gaussian diffusions $z_{t,y}(\tau), t \leq \tau \leq u$, with expectation $m(\tau) = q_{t,y}(\tau)$ and covariance $K_t(\tau_1, \tau_2)$ as above.

Those diffusion processes are more appropriate for semiclassical physics than the Wiener process involved in the conventional Feynman-Kac formula (2.4). In particular, they admit a most welcome regularized concept of derivative (or "drift") which provides the mean forward direction of the movement:

$$Dz_{t,y}(\tau) = \lim_{\Delta\tau \downarrow 0} E_\tau \left[\frac{z_{t,y}(\tau + \Delta\tau) - z_{t,y}(\tau)}{\Delta\tau} \right]$$

$$\equiv B(z_{t,y}(\tau), \tau) \tag{2.11}$$

$$= \dot{F}(\tau) F^{-1}(\tau) z_{t,y}(\tau) + F^{-1}(\tau) \delta$$

where E_τ denotes a conditional expectation given $z_{t,y}(\tau)$ and δ is the constant mentioned in $\mathcal{D}(C^{-1})$. Notice that only $F(.)$ enters in (2.11), namely the solution carrying the future boundary condition (at time u) and not F_t^*.

Suppose now that the conditioning "to be in y at time t" is in the future of the considered time interval, say [s,t]. Then the relevant second variation action functional $\delta^2 \overline{S}$ should involve, instead of Eq. (2.7), a family of initial conditions

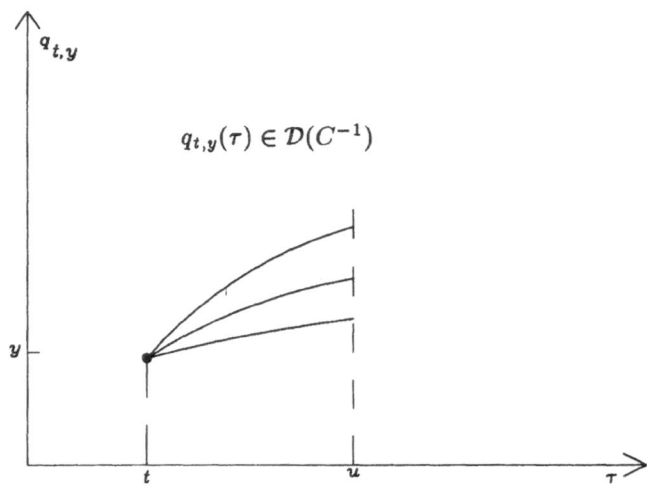

Fig. 1.

corresponding to a domain $\mathcal{D}^*(C^{-1})$ such that

$$
\left\{
\begin{array}{rcl}
\dot{F}_s^* q(s) - F_s^* \dot{q}(s) &=& \delta* \\[2mm]
q(t) &=& y
\end{array}
\right.
\tag{2.12}
$$

for any coefficients with $|F_s^*| + |\dot{F}_s^*| > 0$. $\delta*$ is a coefficient playing here a role analogous to δ.

Consider $\begin{pmatrix} F_{.s}^* \\ \dot{F}_s^* \end{pmatrix}$ as initial boundary condition of the same Jacobi equation (2.9) but now on [s,t]. Denote by $F^*(\tau)$ the associated solution.

This provides another collection of Gaussian diffusions $z^{t,y}(\tau), s \leq \tau \leq t$, such that

$$
D * z^{t,y}(\tau) = \lim_{\Delta\tau \downarrow 0} E_\tau \left[\frac{z^{t,y}(\tau) - z^{t,y}(\tau - \Delta\tau)}{\Delta\tau} \right]
$$

$$
= B * (z^{t,y}(\tau), \tau)
\tag{2.13}
$$

$$
= \dot{F}^*(\tau) F^{*^{-1}}(\tau) z^{t,y}(\tau) - F^{*^{-1}}(\tau)\delta * .
$$

Notice that the conditioning in E_τ is, this time, in the future. The associated integral kernel of the covariance C is now of the symmetric form

$$
K_*^t(\tau_1, \tau_2) =
\left\{
\begin{array}{ll}
F^t(\tau_1) F^*(\tau_2), & s < \tau_2 < \tau_1 < t \\[2mm]
F^t(\tau_2) F^*(\tau_1), & s < \tau_1 < \tau_2 < t
\end{array}
\right.
$$

where $F^t(.)$ is solution of Eq. (2.9), linearly independent of $F^*(.)$ and taking care of the final conditioning $z^{t,y}(t) = y$. The picture is, therefore, the following:

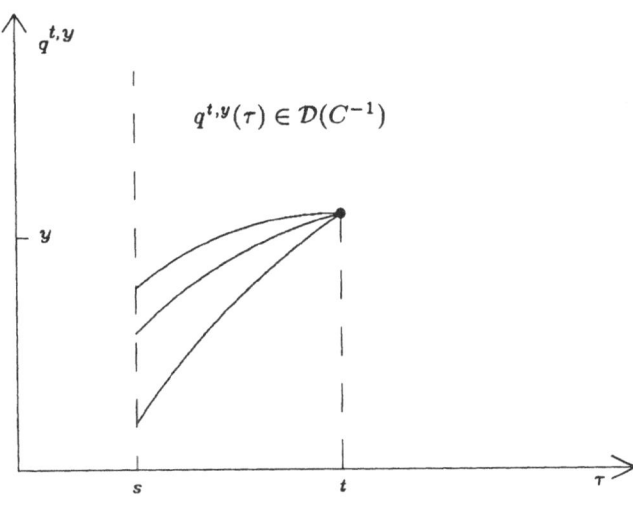

Fig. 2.

As before, there is a Gaussian (or "Normal") diffusion $z^{t,y}(.)$ with mean $q^{t,y}(\tau)$ and covariance kernel K_*^t. In short, $z^{t,y}(\tau) = \mathcal{N}(q^{t,y}(\tau), K_*^t(\tau_1, \tau_2))$.

Each of these two pictures displays an "arrow of time" due to the effect of conditioning, which can be interpreted as a measurement at y.

Although we shall not elaborate here on this interesting physical aspect, let us observe that a concept as simple as the one of conditional expectation does not exist in Quantum Mechanics (because there is no probability measure, in real time, compatible with all what we want in quantum dynamics). On the other hand, it is indeed agreed since N. Bohr that only the measurement process itself bring irreversibility in an otherwise, conservative, quantum system.

Now let's take the two abovementioned pictures together and integrate over the conditioning y at time t: Then, one can show [6] that there is an unique Gaussian diffusion on [s,u], denoted by $z(\tau)$, having both drifts B and B_* as above. In one dimension, it is $z(\tau) = \mathcal{N}(m(\tau), c(\tau))$, where

$$\begin{cases} m(\tau) = W^{-1}(F(\tau)\delta * + F^*(\tau)\delta) \\ C(\tau) = W^{-1}FF_*(\tau) \end{cases} \qquad (2.14)$$

and $W = W(F, F_*) = \dot{F}_*F - F_*\dot{F}$ is the Wronskian of the two abovementioned linearly independent solutions F, F_* of the (homogeneous) Jacobi equation (2.9). Notice that $m(\tau)$ is nothing other than the general solution of (2.9). This implies that the two constant δ, δ_* should be determined by the boundary conditions of the particular classical trajectory $q(\tau), s < \tau < u$ that we wish to quantize.

The product form of $c(\tau)$ (Eq. (2.14)) is hardly new. It is well known that any Gaussian Markovian process has a covariance kernel which is a product. What is special about the Gaussian measures characterized by (2.14) ("Bernstein measures" [3]) is the special relationship between F and F_* when the measure "comes from quantum mechanics". It is necessarily a relation of time reversal. The meaning of this remark will become clearer afterwards.

Let us now assume that the manifold of the boundary conditions associated to Eq. (2.9) for the diffusion $z(\tau)$, namely

$$\dot{F}_s^* q(s) - F_s^* \dot{q}(s) = \delta_*$$

$$F_u \dot{q}(u) - \dot{F}_u q(u) = \delta$$

(2.15)

is invariant under time reversal, as well as the term $\nabla^2 V(z^0(t))$.

It is the case if, under time reversal, $F \leftrightarrow F_*$ and $\delta \leftrightarrow \delta_*$.
Therefore, since the Wronskian itself is invariant under time reversal in this sense, it is clear by inspection of (2.14) that the Gaussian Bernstein measures produced in this way are themselves invariant under time reversal although, generally, not stationary. Notice that the time symmetry of the expectations $m(\tau)$ comes from a linear combination of the time asymmetric solutions F and F_*. These measures belong to a much larger collection of time reversible probability measures associated with Quantum Dynamics (Cf. §3) and have been introduced in 1985-86 on the basis of Schrödinger's idea [13].

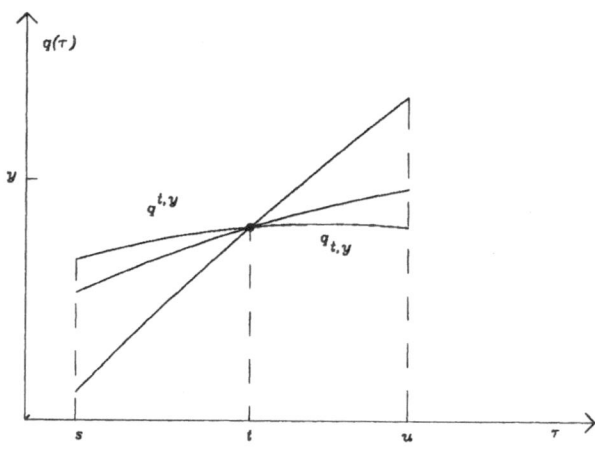

Fig. 3.

Before considering some examples, let us observe that F and F_* are related, according to the classical Sturm-Liouville theory, by

$$F_*(\tau) = F(\tau)[F_*(s) + \int_s^\tau F^{-2}(t)dt.W], \qquad s \leq \tau \leq u$$

and that many unusual properties of the Gaussian Bernstein measures follow from this relation (Cf. [6]).

3. Examples and discussion

Let us consider the classical Jacobi equation with source j associated with $\delta^2 \overline{S}$ as in §2,

$$-\frac{d^2}{dt^2}q(t) + \nabla^2 V(z^0(t))q(t) = -j(t) . \qquad (3.1)$$

For a Bridge, namely a Bernstein diffusion starting from x at time s and ending in z at time $u > s$, the relevant boundary conditions of Eq. (3.1) are clearly those for the two elementary Lagrangian manifolds

$$q(s) = x \qquad \text{and} \qquad q(u) = z . \qquad (3.2)$$

They belong to the general (unmixed) boundary conditions of Sturm-Liouville type:

$$\begin{cases} \beta_s q(s) + \gamma_s \dot{q}(s) = \alpha_s, & |\beta_s| + |\gamma_s| > 0 \\ \\ \beta_u q(u) + \gamma_u \dot{q}(u) = \alpha_u, & |\beta_u| + |\gamma_u| > 0 . \end{cases} \qquad (3.3)$$

Our above experience with the effect of conditioning suggests that $F^*(.)$ satisfies the past homogeneous condition $F_*^* = 0$. Moreover, $\dot{F}_*^* = 1$. Symmetrically, the future boundary condition should be $F_u = 0, \dot{F}_u = -1$.

Since the general solution of the homogeneous Jacobi equation (i.e. $m(\tau)$) should satisfy the given inhomogeneous boundary conditions (3.2) this implies that

$$\delta* = x \qquad \text{and} \qquad \delta = z . \qquad (3.4)$$

For example, for the free case $V = 0$, this provides us with

$$F_*(t) = t - s \qquad \text{and} \qquad F(t) = u - t \qquad (3.5)$$

a pair of solutions reflecting manifestly some time reversal symmetry, already present in the given boundary data.

The two drifts B and B_* of the associated (free) Bernstein Bridge are then (Cf. Eqs. (2.11), (2.13)):

$$B(y,t) = \frac{z-y}{u-t}$$

and   (3.6)

$$B_*(y,t) = \frac{y-x}{t-s}$$

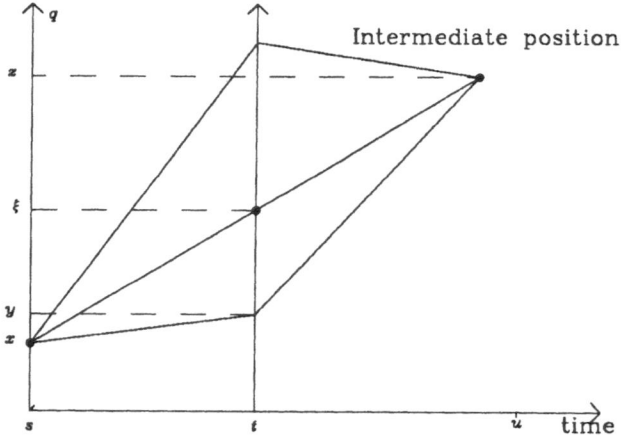

Fig. 4.

Clearly, this can be interpreted as a kind of collection of free "interpolations" between the fixed positions x and z.

Misawa has considered these kind of interpolations [7]. Also notice the special status of $y = \xi$, the only point interpolating between x and z without a discontinuity of the time derivative.

Suppose now that, for the same Jacobi equation (3.1), we select instead of (3.2) the boundary conditions of the usual initial value problem:

$$q(s) = x \qquad \text{and} \qquad \dot{q}(s) = p .$$ (3.7)

The associated constraint on the expectation $m(\tau)$ implies that

$$\delta_* = \dot{F}_s^* x - F_s^* p$$

(3.8)

$$\delta = -\dot{F}_s x + F_s p .$$

Now the initial value boundary data (3.7) are not of the Sturm-Liouville type (3.3). As a consequence, in the classical solution of Eq. (3.1), i.e.

$$q(t) = m(t) + \int K(t, \tau)j(\tau)d\tau$$

the Green kernel K is of the form

$$K_R(t, \tau) = \begin{cases} 0 & s < t < \tau \\ \dfrac{F(\tau)F^*(t) - F(t)F^*(\tau)}{W} & s < \tau < t \,. \end{cases} \quad (3.9)$$

This ("retarded") Green function is not symmetric under time reversal in the abovementioned sense (i.e. $F^* \rightleftarrows F$). So it is immediately clear that the initial value boundary data are not proper for this Semiclassical quantization procedure inspired by Schrödinger. The same is true of the final value problem and its associated ("advanced") Green function K_a : it manifests, as well, an "arrow of time".

It is a nice exercise, in this time–symmetric probabilistic context to reconsider the zoology of Green functions whose extension in infinite dimension is so puzzling otherwise.

For example, the "commutator of positions Green function",

$$Kc(t, \tau) \equiv K_a(t, \tau) - K_R(t, \tau)$$
$$= W^{-1}(F(t)F^*(\tau) - F(\tau)F^*(t)) \quad \text{for any} \quad t, \tau, \quad (3.10)$$

is not symmetric under $F \rightleftarrows F_*$ but deserves, indeed, its name since

$$\frac{\partial}{\partial t}Kc(t, \tau)|_{t=\tau} = -1$$

corresponds precisely to

$$W^{-1}(F(t)\dot{F}^*(t) - F^*(t)\dot{F}(t)) = 1 \quad (3.11)$$

so that we are entitled to interpret it as an ordinary differential equation counterpart of the (one dimensional) Heisenberg commutation relation between position and momentum.

All such elementary Green functions can be reinterpreted directly in terms of the pair $(F(t), F^*(t))$ used in Eq. (2.14) to construct any Gaussian Bernstein measures.

Let us stress that what is causal (in the naïve sense) is not necessarily relevant for our probabilistic (Semiclassical of full) quantization. For example, the boundary conditions (3.7) seem, classically, the most natural ones but, nevertheless, break the fundamental time reversibility of Bernstein measures. This should be kept in mind also in infinite dimension.

One can show that the abovementioned (Euclidean) strategy of construction of Gaussian measures associated with a given Lagrangian $\overline{L}$ exhausts the semiclassical probabilistic content of this physical system. Actually, there is a one to one correspondence between the underlying classical trajectories and the associated Gaussian measures [14-6].

Now Gaussian Bernstein measures, in our approach, are not produced via the traditional Feynman-Kac argument of perturbation of the Gaussian case. They relate to the existence of a fully time-symmetric probabilistic interpretation of the heat equation (2.2), analogous to the one for which Max Born (to whom the present Symposium is dedicated) went into History for the Schrödinger equation.

Let $H = - \frac{\hbar^2}{2} \triangle + V$ be the Hamiltonian observable corresponding to $\overline{L}$. Then, to the quantum probabilistic interpretation of Born, namely

$$\int_B \psi\overline{\psi}(x,t)dx = Pr\,\{\text{System} \in B \text{ at time } t\} \tag{3.12}$$

corresponds in Euclidean Quantum Mechanics [14]

$$\int_B \eta_*\eta(x,t)dx = Pr\,\{\text{System} \in B \text{ at time } t\} \tag{3.13}$$

where - $\hbar\dot{\eta}_* = H\eta_*$ and $\hbar\dot{\eta} = H\eta$. The relation between η and η_* is a relation of time reversal replacing, in the Euclidean world, the complex conjugation involved in (3.12). What we have done here is to describe the semiclassical limit of this Euclidean quantization method due, in essence, to E. Schrödinger [13].

The original problem of Schrödinger (How to produce diffusions from the data of two probabilities at different times?) was found natural much later, in a real time probabilistic approach to quantum mechanics, by K. Yasue [8]. The dynamical structure of the two resulting probabilistic theories (Nelson's "Stochastic Mechanics" on one hand and Euclidean Quantum mechanics on the other are, however, quite distinct [6, 11]). In particular, the Euclidean framework preserves the relation with the multitime predictions of quantum theory, which is lost in the real time framework, even in the Gaussian case.

Another recent probabilistic representation of quantum dynamics in term of complex Markov processes is due to Haba [9].

P. Garbaczewski has published a number of papers in relation to Bernstein diffusions. Cf. for example [10].

By construction, $\eta\eta_*(x,t).dx = p(x,t).dx$ is the probability density of the Bernstein diffusions $z(t) = z^\hbar(t)$, for t in a certain interval, processes which are generally not Gaussian.

What we have done here amounts to consider the two first terms in the (rigorously defined) $\hbar$-expansion:

$$z^\hbar(t) = z^0(t) + \sqrt{\hbar}z^{Sc}(t) + \hbar z^{(2)}(t) + \dots \tag{3.14}$$

Up to the order $\hbar$ this diffusion is a Gauss-Markov process with linear and time dependent drifts, given by Eqs. (2.11) and (2.13) Cf. [6].

The underlying rigorous calculus of functionals of Bernstein diffusions uses Malliavin Calculus: Cf. [11].

The abovementioned construction of Gaussian (and non Gaussian) measures does not depend on the dimension. For some steps in infinite dimension, cf [12].

In conclusion, one may observe that, even in the simplest possible context of Quantum Physics, the semiclassical one, the role of time reversal appears to be both more subtle and fundamental than generally acknowledged.

Nevertheless, it should be also said that "Euclidean Quantum mechanics" will really deserve its designation only when it will show that the Euclidean detour is more than a technical trick, and suggests both problems and solutions not yet accessible to orthodox quantum mechanics. This should be true, in particular, in the semiclassical regime and the fascinating problem associated with the quantization of classical chaotic systems. For this, however, it is possible that quantum mechanics as we know it has to be generalized. In this case, we share Feynman's feeling [2, p 173] that probabilistic approaches may show us the way.

Acknowledgment

The author is grateful to the organizing committee for the invitation to the Third Max Born Symposium.

References

1. P. Martin-Löf, Journal of Inform. and Control 9, 602 (1966).
2. R. Feynman and A. Hibbs, "Quantum Mechanics and Path integrals", McGraw-Hill, N.Y. (1965); J. A. Wheeler and R. Feynman, Rev. Mod. Phys. 17, 157 (1945) and 21, 425 (1949).
3. J.C. Zambrini, J. Math. Phys. 27, 2307 (1986); Phys. Rev.A 359, 3631 (1987); Proceed. of IX IAMP Congress, Swansea Ed. B. Simon, A. Truman, I. M. Davies, Hilger, Bristol, 260 (1989).
4. M. Kac, "On some connections between probability theory and differential and integral equations", Proc. of 2nd Berkeley Symposium on Probability and Statistics, Ed. J. Neyman, Univ. California Press Berkeley (1951).
5. B. Simon, "Functional Integration and Quantum Physics", Acad. Press, N.Y. (1979), p 197-198.
6. T. Kolsrud, J.C. Zambrini, "An introduction to the semiclassical limit of Euclidean Quantum Mechanics", J. Math. Phy. 33(4), April 93, 1301.
7. T. Misawa, J. Math. Phys. 34, 775 (1993).
8. K. Yasue, J. Math. Phys. 22, 1010 (1981).
9. Z. Haba, Phys. Lett. A 175, 371 (1993).
10. P. Garbaczewski, J.P. Vigier, Phys. Lett. A 167, 445 (1992).
11. A.B. Cruzeiro and J.C. Zambrini, J. Funct. Anal. 96(1), 62 (1991); "Feynman's Functional Calculus and Stochastic Calculus of Variations" in "Stochastic Analysis and applications", Progr. in Prob. Series vol 26, Ed. A.B. Cruzeiro and J.C. Zambrini, Birkhäuser, Boston (1991).
12. A.B. Cruzeiro and J.C. Zambrini, "Ornstein-Unlenbeck Processes as Bernstein processes", in "Barcelona Seminar on Stochastic Analysis", Edit D. Nualart and Marta Sanz Solé, Birkhäuser, Progress in Prob. Series n⁰ 32, p. 40 (1993); A.B. Cruzeiro, Z. Haba and J.C. Zambrini, "Bernstein diffusions and Euclidean Quantum Field Theory", to appear in a Birkhäuser volume, Monte Verita Meeting, Ascona (1993).
13. E. Schrödinger, Ann. Inst. H. Poincaré, 2, 269 (1932).
14. S. Albeverio, K. Yasue and J.C. Zambrini, Ann. Inst. H. Poincaré, Phys. Th. 49(3), 259 (1989).

STOCHASTIC RESONANCE IN BISTABLE SYSTEMS WITH FLUCTUATING BARRIERS

L. GAMMAITONI[1,2], F. MARCHESONI[2,3], ,
E. MENICHELLA-SAETTA[1] and S. SANTUCCI[1]
[1] *Dipartimento di Fisica, Universita' di Perugia, I-06100 PERUGIA (Italy)*
[2] *Istituto Nazionale di Fisica Nucleare, Sezione di Perugia, I-06100 PERUGIA (Italy)*
[3] *Dipartimento di Matematica e Fisica, Universita' di Camerino, I-62032 CAMERINO (Italy)*

Abstract. A multiplicative bistable system perturbed by a periodic forcing term is shown to exhibit stochastic resonance with increasing the intensity of the multiplicative noise. Such an effect is related to the phenomenon of stochastic stabilization, which takes place in the unperturbed system.

1. The amplitude of the periodic component of the output signal from a periodically modulated bistable system in the presence of *additive* external noise, increases with the noise intensity up to a maximum value when the forcing frequency approaches the rate of the noise-induced switch process. Such a phenomenon, termed *stochastic resonance* [1] (SR), was investigated by a number of authors [1-8] and found application in many areas of natural sciences [9].

There exist many cases of physical interest [10] where the role of fluctuating control parameter is played by a *multiplicative* noise [10-13]. In the present report we show that SR may be caused by rising the multiplicative noise intensity, as well, thus resembling the additive case addressed in the current literature [9]. In this case, however, the enhancement of the periodic component of the output signal is to be traced back to the phenomenon of *stochastic stabilization*, which would occur in the absence of periodic modulation [10-11].

In our investigation, multiplicative SR was detected by analogue simulating the overdamped bistable system described by the following stochastic differential equation [14]

$$\dot{x} = -V'(x) + x\xi(t) + \eta(t) + A\cos\omega_0 t \tag{1}$$

with

$$V(x) = -\frac{a}{2}x^2 + \frac{b}{4}x^4 \qquad (a, b > 0) \tag{2}$$

The fluctuating parameters $\xi(t)$ and $\eta(t)$ are stationary, uncorrelated, zero-mean valued, gaussian random processes with autocorrelation functions

$$< \xi(t)\xi(0) >= 2Q\delta(t) \tag{3}$$

and

$$< \eta(t)\eta(0) >= 2D\delta(t) \tag{4}$$

The origin of time in eq. (1) has been fixed arbitrarily. Furthermore, the amplitude of the perturbing term $A\cos\omega_0 t$ is assumed to be so small that the bistable nature of

Z. Haba et al. (eds.), Stochasticity and Quantum Chaos, 209–216.
© 1995 Kluwer Academic Publishers.

the process $x(t)$ is retained, i.e. $|A|x_0 \ll \Delta V$ with $x_0 = (a/b)^{1/2}$ and $\Delta V = a^2/4b$. Here $\pm x_0$ denote the stable minima and ΔV the barrier height of the potential (2).

2. The phenomenon of stochastic stabilization was predicted originally for the purely multiplicative stochastic process described by eq. (1) with $\eta(t) = 0$ (or $D = 0$) and $A = 0$. The corresponding probability distribution function,

$$P_0(x) = N_0|x|^{-1+2k} exp\left(-k\frac{x^2}{x_0^2}\right) \tag{5.a}$$

with $N_0 = 2(k/x_0^2)^k/\Gamma(k)$ and $k = a/2Q$, exhibits two peculiar properties:
(i) the process $x(t)$ is confined on one half-axis (either positive or negative) at any time, depending on the initial conditions;
(ii) a sort of noise-induced phase transition [10] occurs at $Q = a$. For weak noise intensity $P_0(x)$ peaks at $x_0(1 - 2k)^{1/2}$. On increasing Q above a, instead, $P_0(x)$ becomes singular (but still normalizable) at the origin. This effect, termed stochastic stabilization, is characterized by the appearance of long tails in $P_0(x)$, i.e. by large high-order moments $< x_n(Q) >= (x_0^2/k)^{n/2}\Gamma(k + n/2)/\Gamma(k)$. Note that, anyway, $< x^2(Q) >= x_0^2$ and $< x(Q) >$ tends to zero monotonically with increasing Q.
[On passing we also note that by adding an additive noise [10], no matter how weak, $x(t)$ diffuses on the entire x-axis. The probability distribution is then symmetric for $x \rightarrow -x$,

$$P_0(x) = N_0'\left(x^2 + \frac{D}{Q}\right)^{-\frac{1}{2}+k\left(1+\frac{D}{x_0^2 V}\right)} exp\left(-k\frac{x^2}{x_0^2}\right) \tag{5.b}$$

where N_0' is an appropriate normalization constant. For weak additive noise, $D \ll x_0^2 Q$, the distribution (5.b) comes fairly close to (5.a) (with $N_0' = N_0/2$), apart from a small symmetric neighbourhood around the origin, where it stays finite.]

In the presence of a *static* tilting of the potential $V(x)$, i.e. $A \neq 0$ and $\omega_0 = 0$, the behaviour described in (ii) gets modified. The relevant probability distribution function $P_0(x, A)$ is easily obtained by solving the corresponding Fokker-Planck equation in the stationary limit $t \rightarrow \infty$ [10]:

$$P_0(x, A) = N_0(A)|x|^{-1+2k} exp\left(-k\frac{x^2}{x_0^2} - \frac{A}{Q|x|}\right) \tag{6}$$

with $N_0(A)$ a suitable normalization factor. $P_0(x, A)$ vanishes on the negative half-axis for $A > 0$ and viceversa. Since $P_0(0, A) = 0$ for $A \neq 0$ we expect $P_0(x, A)$ to show one peak located at $x = \pm x_m$ with

$$x_m(Q) \simeq \begin{cases} x_0 + \dfrac{1}{2}\dfrac{|A|}{a - Q} & Q \ll a \\[3mm] \left(\dfrac{|A|}{b}\right)^{\frac{1}{3}} & Q = a \\[3mm] \dfrac{|A|}{Q - a} & Q \gg a \end{cases} \tag{7}$$

The signs $\pm$ refer to the conditions $\pm A > 0$, respectively.

Let us assume, now, that $V(x)$ is tilted *periodically* in time, like in eq. (1) with $A > 0$ and $\omega_0 > 0$. The process $x(t)$ is no longer stationary and a time dependent probability distribution function $P(x, A; t)$ is required, in principle, to describe its steady state. For the purpose of our analysis the time average of $P(x, A; t)$ over one modulation cycle $T_0 = 2\pi/\omega_0$ will suffice. In the limit of low- frequency periodic modulation discussed below, i.e. for $\omega_0 \ll a$, we evaluated such an average in the adiabatic approximation [3,8],

$$P_a(x) = \frac{1}{T_0} \int_0^{T_0} P_0(x, A(t)) \, dt \qquad (8)$$

Here $P(x, A; t)$ is approximated to $P_0(x, A(t))$, i.e. to the distribution of eq. (6), where A has been replaced by $A(t) = A\cos \omega_0 t$. In fig. 1 we report $P_a(x)$ (result of analogue simulation) at increasing values of Q. The agreement with theoretical prediction is fairly good.

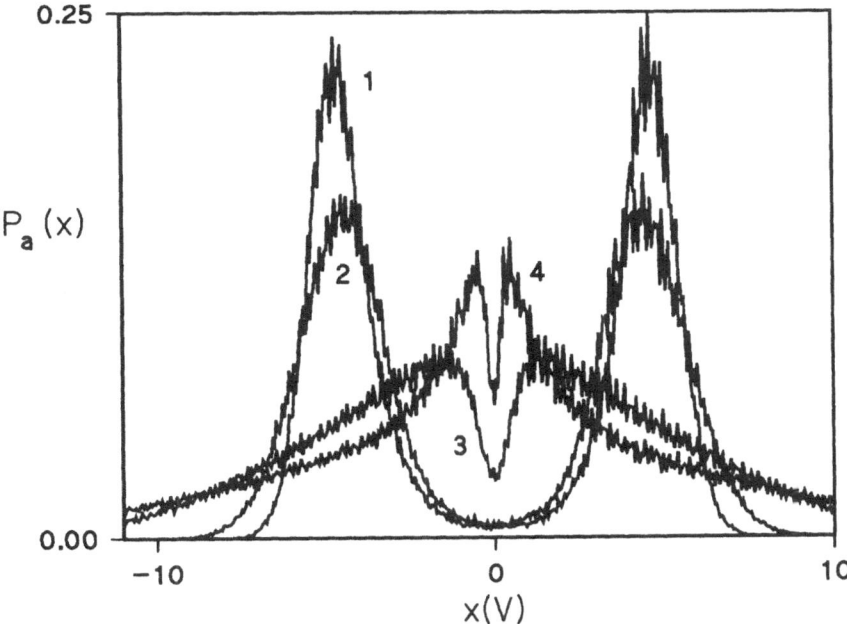

Fig. 1. $P_a(x)$ *versus* x for different values of Q. 1, $Q = 691 \ s^{-1}$. 2, $Q = 1920 \ s^{-1}$. 3, $Q = 11570 \ s^{-1}$. 4, $Q = 20060 \ s^{-1}$. $x_0 = 2.2 \ V$, $a = 10^4 \ s^{-1}$, $A = 0.5ax_0$, $\nu_0 = \omega_0/2\pi = 30 \ Hz$.

3. In order to interpret the results of our simulation work we recall two further characteristic properties of multiplicative processes. When simulating a purely multiplicative bistable system by means of an analogue device two limitations are unavoidable:

(iii) the presence of small additive noise $(D > 0)$ due to ubiquitous fluctuations in the circuitry. As a consequence our results for $\eta = 0$ are to be taken with some

caution. Any attempt at a full analytical treatment of eq. (1) with both $D > 0$ and $Q > 0$ proved unsuccessful [15]. Numerical algorithms based on continued fraction expansions are the only tools available to date to crack down the problem [12,13]. In ref. 12 the escape rate μ between the minima of $V(x)$, $\pm x_0$, was computed as a function of the multiplicative noise intensity, $\mu = \mu(Q)$. For a given value of the additive noise intensity D, the function $\mu(Q)$ grows linearly with increasing Q from zero (where $\mu(0)$ coincides with the well-known Kramers rate $\mu_k(D)$) up to a critical value $Q = a/4$ (continuum spectrum threshold [10]) and, then, flattens out until the stochastic stabilization condition $Q = a$ is reached. For larger values of Q, $\mu(Q)$ appears to diverge faster than exponentially.

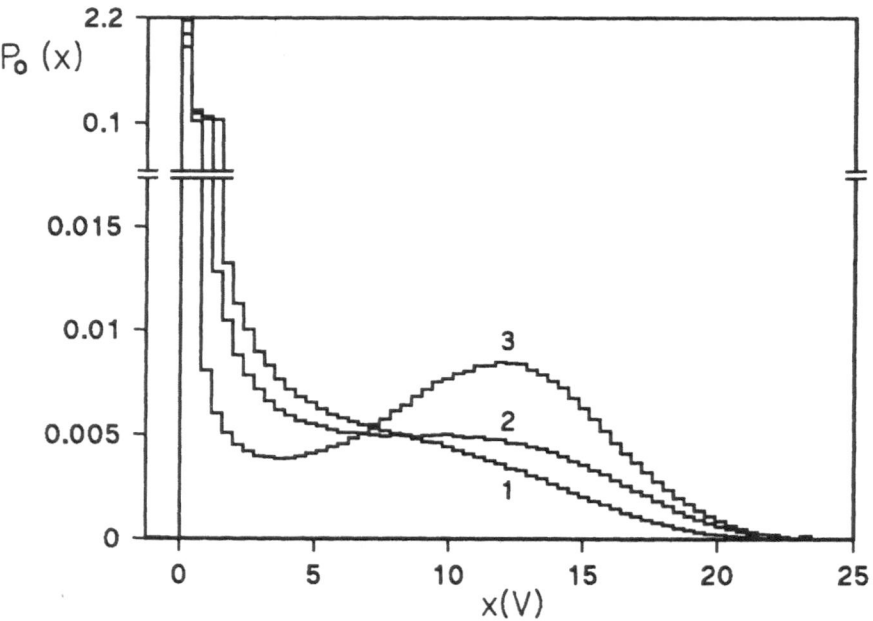

Fig. 2. $P_0(x)$ *versus* x for the same Q and different values of τ_ξ. $Q = 32000\ s^{-1}$, $A = 0$. 1, $\tau_\xi = 5\ 10^{-6}s$. 2, $\tau_\xi = 50\ 10^{-6}s$.

(iv) any commercial noise generator is characterized by a finite correlation time (colored noise). For instance, the typical autocorrelation function of the simulated multiplicative noise $\xi(t)$ is well approximated by an exponential function

$$< \xi(t)\xi(0) > = \frac{Q}{\tau_\xi}\ exp\left(-\frac{|t|}{\tau_\xi}\right) \tag{9}$$

which tends to eq. (3) for vanishingly small values of τ_ξ, only. Effects due to the finite correlation time of $\xi(t)$ are negligible when $1/\tau_\xi$ is larger than any other dynamical frequency in the problem under study. In the purely multiplicative case, $D = 0$, this amounts to requiring that

$$(a + Q)\tau_\xi \ll 1 \tag{10}$$

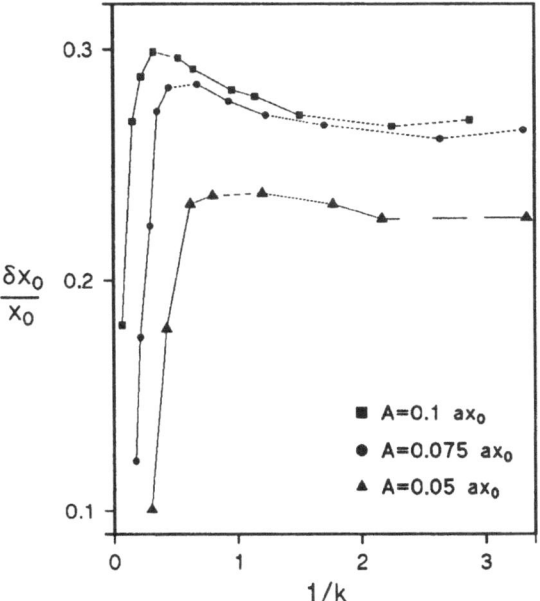

Fig. 3. δx_0 *versus* Q for $D = 0$ and different values of A. Potential parameters as in Fig. 1.

It follows immediately that in the regime of stochastic stabilization, coloured noise effects may show up when increasing Q well beyond the critical value $Q = a$. Under such circumstances, the relevant probability distribution function $P_0(x)$ (5.a) still diverges at $x = 0$, as expected in the white noise case. Most notably, a side peak grows out of the long tail of $P_0(x)$ [13]. We verified that for large τ_ξ values the position of such a peak is well approximated by $x_0 \left(1 + \sqrt{Q/a^2 \tau_\xi}\right)^{1/2}$. This prediction follows the approach to strongly coloured bistable-flow dynamics outlined in ref. 15. Thus, contrary to the white noise case, the second moment of $P_0(x)$ becomes a slowly increasing function of Q with $< x^2(Q) > > x_0^2$. In fig. 2 the Q dependence of $P_0(x)$ has been determined by analogue simulation. The results obtained are consistent with our theoretical predictions.

4. We are now in the position to explain qualitatively the phenomenon of multiplicative SR in the forced bistable system (1) with $\omega_0 \ll a$. Let us consider first the case with $\eta(t) = 0$. Here the experimental evidence for multiplicative SR is limited by the performance of our simulator - see (iii). It is clear from the discussion in sect. 2 that for $D = 0$ the forcing term $A(t)$ alone is responsible for $x(t)$ to switch back and forth between the positive and negative half-axes. Should the adiabatic approximation hold good for any value of Q, the process $x(t)$ would relax

instantaneously towards the stable quasi-equilibrium state described by $P_0(x, A(t))$. Under such circumstances the amplitude $\delta x = \delta x(D, Q)$ of the periodic component of $x(t)$

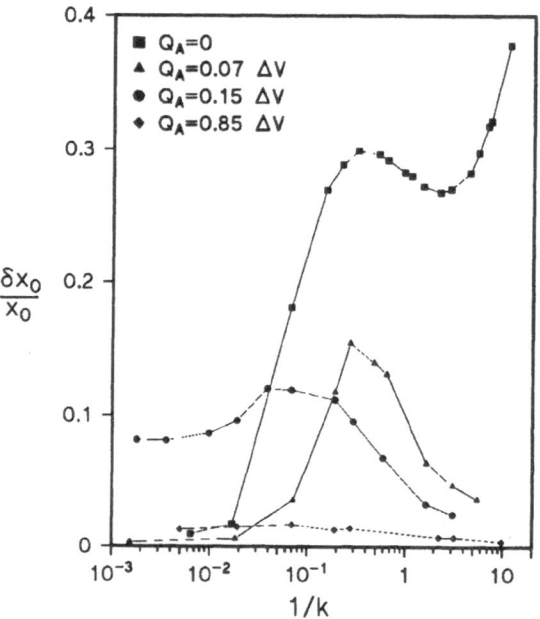

Fig. 4. δx_0 *versus* Q for $A = 0.1ax_0$ and different values of D. Potential parameters as in Fig. 1.

$$< x(t) >= \delta x \, cos(\omega_0 t + \phi) \tag{11}$$

would be determined by

$$\delta x_0(Q) \equiv \delta x(0, Q) \simeq \int_0^\infty P_a(x)x\,dx \tag{12}$$

For the sake of comparison, we notice that the integral on the rhs of eq. (12) is of the order $x_m(Q)$, eq. (7), i.e. a monotonically decreasing function of Q. However, our data for $\delta x_0(Q)$ (fig. 3) deviate from the predicted law (12) both for very small and very large Q-values, thus exhibiting a typical SR behaviour. For very large values of Q the color effects become dominant to the point that the expected decrease of $\delta x_0(Q)$ may become hardly observable. Only in the presence of a rather strong forcing (topmost curve), the predicted decay of $\delta x_0(Q)$ becomes apparent. The dramatic drop of $\delta x_0(Q)$ as Q tends to zero, instead, is due to the fact that in a bistable system the adiabatic approximation may fail even for $\omega_0 \ll a$. The escape time of $x(t)$ out of the half x-axis made unstable by reversing sign of $A(t)$, can be calculated explicitly. For instance, on assuming that x is trapped on the unstable half-axis $x(0) < x_A \simeq A/a$ with $A > 0$, the mean first-passage time τ_A required by

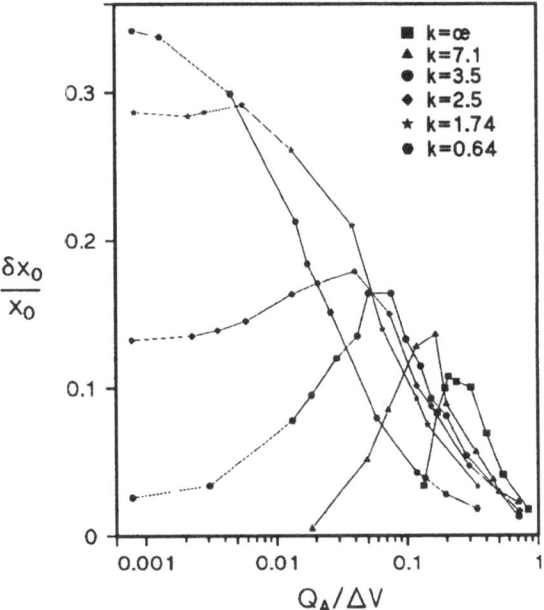

Fig. 5. δx_0 *versus* D for $A = 0.1ax_0$ and different values of Q. Potential parameters as in Fig. 1.

x to escape through x_A onto the stable half-axis $x > x_A$ can be easily calculated at the leading order in $(x_0/x_A)^2$, i.e.

$$\tau_A(Q) = \frac{\Gamma(k)}{2ak^k} \left(\frac{x_0}{x_A}\right)^{2k} \tag{13}$$

The adiabatic approximation holds good for values of Q large enough to guarantee that $\omega_0\tau_A \ll 1$. The time τ_A is a strongly divergent quantity for $x_A/x_0 \to 0$. On increasing Q close to the condition of stochastic stabilization, more precisely for $k \leq 1$, such a divergence is substantially weakened, so that the condition $\omega_0\tau_A(Q) = 1$ is verified for a certain value $Q = Q_A$ and $\delta x_0(Q)$ attains its expected value (12) exponentially for $Q \gg Q_A$ (modulated *inter-well* dynamics). In the opposite limit $\omega_0\tau_A \gg 1$ (or $Q < Q_A$), the steady-state signal $x(t)$ is incoherently distributed over the entire x-axis with oscillating most-probable values at $\pm x_0 + A(t)/2(a - Q)$, see eq. (7), whence $\delta x_0(0) \simeq A/2a$ (modulated *intra-well* dynamics). We remark that the assumption $Ax_0 \ll \Delta V$ introduced in sect. 1 implies that $\delta x_0(0) < \delta x_0(a)$ with $\delta x_0(a) \simeq x_m(a) = (A/b)^{1/3}$. In conclusion, the transition from the intra-well to the inter-well modulated dynamics is the basic mechanism responsible for multiplicative SR.

The phenomenon of multiplicative SR becomes more apparent in our simulations by switching on the additive noise $\eta(t)$. In fig. 4 plots of $\delta x_D(Q) \equiv \delta x(D, Q)$ for several values of D are displayed in order to illustrate the interplay of additive and multiplicative noise. The switch process between the positive and negative half-axes in eq. (1) is now assisted by the additive noise $\eta(t)$, which makes the relevant (finite) switch-time τ_A decrease with D. As a consequence, $\delta x_D(Q)$ rises from zero up to its maximum value for smaller values of Q, compared with $\delta x_0(Q)$. Furthermore, in the presence of additive noise $x(t)$ diffuses onto the unstable half-axis even in the adiabatic limit [12], i.e. for large values of Q, provided that $\mu_k(D)\tau_A \gg 1$. It follows that with increasing D the asymptotic value of $\delta x_D(Q)$ for $Q \to \infty$, decreases much below the corresponding value for $D = 0$, (12). The two competing trends determine a rather broad peak of $\delta x_D(Q)$ centered around the Q value that satisfies the condition $\omega_0\tau_A \simeq 1$ (multiplicative SR). On further increasing D the SR peak is washed out.

Of course, eq. (1) also exhibits additive SR as shown in fig. 5, where the amplitude $\delta x_Q(D) \equiv \delta x(D, Q)$ of the periodic component of $x(t)$ is plotted as a function of D at different values of Q. It is well known [1-8] that at $Q = 0$ the response amplitude $\delta x_0(D)$ peaks for noise intensities D such that $\mu_k(D) \simeq \omega_0$. Therefore, one expects that on increasing Q the resonance condition may be recast as $\mu(Q) \simeq \omega_0$, with $\mu(Q)$ plotted in ref. 12. On making use of the fact that $\mu(Q)$ grows linearly (or slower) with Q in the range $(0, a)$, it is no surprise that, correspondingly, the position of the resonance peak in fig. 5 appears to shift to lower values of D. Finally, the condition is reached that $\delta x_Q(D)$ becomes a monotonic, decreasing function of D. That occurs for values of Q such that $\delta x_D(Q)$ approaches its adiabatic limit (12).

References

1. R. Benzi, G. Parisi, A Sutera and A. Vulpiani, Tellus **34** 10 (1982) and SIAM J. Appl. Math. **43** 565 (1983); C. Nicolis, Tellus **34** 1 (1982)
2. B. McNamara, K. Wiesenfeld and R. Roy, Phys. Rev. Lett. **60** 2628 (1988)
3. L. Gammaitoni, F. Marchesoni, E. Menichella-Saetta and S. Santucci, Phys. Rev. Lett. **62** 349 (1989)
4. P. Jung and P. Hanggi, Europhys. Lett. **8** 505 (1989), Phys. Rev. **A41** 2977 (1990) and **44** 8032 (1991)
5. C. Presilla, F. Marchesoni and L. Gammaitoni, Phys. Rev. **A40** 2105 (1989)
6. H. Gang, G. Nicolis and C. Nicolis, Phys. Rev. **A42** 2030 (1990) and Phys. Lett. **A151** 139 (1990)
7. L. Gammaitoni, F. Marchesoni, M. Martinelli, L. Pardi and S. Santucci, Phys. Lett. **A158** 449 (1991)
8. T. Zhou, F. Moss and P. Jung, Phys. Rev. **A42** 3161 (1990)
9. for a review, see the Proceedings of the NATO Advanced Research Workshop on 'Stochastic Resonance' March 30 - April 3, 1992, San Diego (Ca), J. Stat. Phys. **70** (1/2) (1993)
10. A. Schenzle and H. Brand, Phys. Rev. **A20** 1628 (1979)
11. R. Graham and A. Schenzle, Phys. Rev. **A25** 1731 (1982)
12. S. Faetti, P. Grigolini and F. Marchesoni, Z. Phys. **B47** 353 (1982)
13. P. Jung and H. Risken, Phys. Lett. **A103** 38 (1984)
14. F. Marchesoni, E. Menichella-Saetta, M. Pochini and S. Santucci, Phys. Rev. **A37** 3058 (1987)
15. P. Hanggi, P. Jung and F. Marchesoni, J. Stat. Phys. **54** 1367 (1989) and references therein.

Index

217

LIST OF PARTICIPANTS

1. N.L. Balazs
   Department of Physics
   State University of New York
   STONY BROOK NY 11794
   USA

2. Ph. Blanchard
   Institut für Theoretische Physik
   Universität Bielefeld
   D-4800 BIELEFELD 1
   GERMANY

3. Yu.L. Bolotin
   Kharkov Institute of Physics and Technology
   1 Akademicheskaya St.
   310108 KHARKOV
   UKRAINE

4. G. Casati
   Dipartimento di Fisica
   Universita degli Studi di Milano
   Via Celoria 16
   I-20133 MILANO
   ITALY

5. W. Cegła
   Institute of Theoretical Physics
   University of Wrocław
   Pl. M. Borna 9
   WROCLAW
   POLAND

6. T. Dittrich
   Institut für Physik
   Universität Augsburg
   Memminger Strasse 6
   D-8900 AUGSBURG
   GERMANY

7. B. Eckhardt
   Fachbereich Physik und ICBM
   C.v. Ossietzky Universität
   2900 OLDENBURG
   GERMANY

8. P. Garbaczewski
   Institute of Theoretical Physics
   University of Wrocław
   Pl. M. Borna 9
   WROCLAW
   POLAND

9. P. Gaspard
   Faculte des Sciences
   de l'Universite Libre de Bruxelles
   Campus Plaine CP 226
   Boulevard du Triomphe
   1050 BRUXELLES
   BELGIUM

10. N. Gisin
    Group of Applied Physics
    University of Geneva
    20, rue de l'Ecole-de-Medecine
    CH-1211 GENEVE 4
    SWITZERLAND

11. P. Gusin
    Institute of Theoretical Physics
    University of Wrocław
    Pl. M. Borna 9
    WROCLAW
    POLAND

12. Z. Haba
    Institute of Theoretical Physics
    University of Wrocław
    Pl. M. Borna 9
    WROCLAW
    POLAND

13. W. Hann
    Institute of Theoretical Physics
    University of Wrocław

Pl. M. Borna 9
WROCLAW
POLAND

14. W.D. Heiss
    Center for Nonlinear Studies
    and Department of Physics
    University of the Witwatersrand
    P.O. Wits 2050
    JOHANNESBOURG
    SOUTH AFRICA

15. H. Herzel
    Sektion Physik
    Humbold-Universität
    Invalidenstrasse 42
    0-1040 BERLIN
    GERMANY

16. J. Hietarinta
    Wihuri Physical Laboratory
    and Department of Physical Sciences
    University of Turku
    SF-20 500 TURKU
    FINLAND

17. L. Jakóbczyk
    Institute of Theoretical Physics
    University of Wrocław
    Pl. M. Borna 9
    WROCLAW
    POLAND

18. B. Jancewicz
    Institute of Theoretical Physics
    University of Wrocław
    Pl. M. Borna 9
    WROCLAW
    POLAND

19. H.R. Jauslin
    Dep. Physique – Labo SMIL
    Universite de Bourgogne
    6, bd Gabriel
    21100 DIJON
    FRANCE

20. N.G. van Kampen
    Instituut voor Theoretische Fysica
    Rijksuniversiteit te Utrecht
    UTRECHT
    NETHERLANDS

21. W. Karwowski
    Institute of Theoretical Physics
    University of Wrocław
    Pl. M. Borna 9
    WROCLAW
    POLAND

22. W. Kirsch
    Institut für Mathematik
    Ruhr-Universität Bochum
    D-4630 BOCHUM 1
    GERMANY

23. A. Knauf
    Technische Universitat
    Fachberiech 3 - Mathematik
    Strasse des 17 Juni
    D - 1000 BERLIN 12
    GERMANY

24. G. Kondrat
    Institute of Theoretical Physics
    University of Wrocław
    Pl. M. Borna 9
    WROCLAW
    POLAND

25. P. Leboeuf
    Division de Physique Theorique
    Institut de Physique Nucleare
    91406 ORSAY
    FRANCE

26. J. Lukierski
    Institute of Theoretical Physics
    University of Wrocław
    Pl. M. Borna 9
    WROCLAW
    POLAND

27. F. Marchesoni
    Dipartimento di Fisica
    Universita di Perugia
    I - 06100 PERUGIA
    ITALIA

28. L. Molinari
    Dipartimento di Fisica
    Universita degli Studi di Milano
    Via Celoria 16
    I-20133 MILANO
    ITALY

29. M. Mystkowski
    Institute of Theoretical Physics
    University of Wrocław
    Pl. M. Borna 9
    WROCLAW
    POLAND

30. Z. Popowicz
    Institute of Theoretical Physics
    University of Wrocław
    Pl. M. Borna 9
    WROCLAW
    POLAND

31. K. Rapcewicz
    Institute of Theoretical Physics
    University of Wrocław
    Pl. M. Borna 9
    WROCLAW
    POLAND

32. D. Richards
    Mathematics Department
    Open University
    MILTON KEYNES
    MK 7 6AA, UK

33. L. Schimansky - Geier
    Sektion Physik
    Humbold-Universität
    Invalidenstrasse 42
    0-1040 BERLIN

GERMANY

34. Yu.P. Virchenko
    Kharkov Institute of Physics and Technology
    1 Akademicheskaya St.
    310108 KHARKOV
    UKRAINE

35. A. Vulpiani
    Dipartimento di Fisica
    Universita di Roma "La Sapienza"
    Piazzale Aldo Moro 2
    I - 00185 ROMA
    ITALIA

36. M. Wolf
    Institute of Theoretical Physics
    University of Wrocław
    Pl. M. Borna 9
    WROCLAW
    POLAND

37. J.C. Zambrini
    Centro de Fisica da Materia
    Condensada
    Av. Gama Pinto, 2
    1699 LISBOA
    PORTUGAL

38. K. Życzkowski
    Institut of Physics
    Jagiellonian University
    Reymonta 4
    KRAKÓW
    POLAND

39. K. Yasue
    Notre Dame Seishin University
    OKAYAMA 700
    JAPAN

Other *Mathematics and Its Applications* titles of interest:

F. Langouche, D. Roekaerts and E. Tirapegui: *Functional Integration and Semiclassical Expansions*. 1982, 328 pp. ISBN 90-277-1472-X

N.E. Hurt: *Geometric Quantization in Action. Applications of Harmonic Analysis in Quantum Statistical Mechanics and Quantum Field Theory*. 1982, 352 pp.
ISBN 90-277-1426-6

C.P. Bruter, A. Aragnal and A. Lichnerowicz (eds.): *Bifurcation Theory, Mechanics and Physics. Mathematical Developments and Applications*. 1983 400 pp. *out of print*, ISBN 90-277-1631-5

N.H. Ibragimov: *Transformation Groups Applied to Mathematical Physics*. 1984, 414 pp. ISBN 90-277-1847-4

V. Komkov: *Variational Principles of Continuum Mechanics with Engineering Applications. Volume 1: Critical Points Theory*. 1986, 398 pp.
ISBN 90-277-2157-2

R.L. Dobrushin (ed.): *Mathematical Problems of Statistical Mechanics and Dynamics. A collection of Surveys*. 1986, 276 pp. ISBN 90-277-2183-1

P. Kree and C. Soize: *Mathematics of Random Phenomena. Random Vibrations of Mechanical Structures*. 1986, 456 pp. ISBN 90-277-2355-9

H. Triebel: *Analysis and Mathematical Physics*. 1987, 484 pp.
ISBN 90-277-2077-0

P. Libermann and Ch.M. Marle: *Sympletic Geometry and Analytical Mechanics*. 1987, 544 pp. ISBN 90-277-2438-5 (hb), ISBN 90-277-2439-3 (pb).

E. Tirapegui and D. Villarroel (eds.): *Instabilities and Nonequilibrium Structures*. 1987, 352 pp. ISBN 90-277-2420-2

W.I. Fushchich and A.G. Nikitin: *Symmetries of Maxwell's Equations*. 1987, 228 pp. ISBN 90-277-2320-6

V. Komkov: *Variational Principles of Continuum Mechanics with Engineering Applications. Volume 2: Introduction to Optimal Design Theory*. 1988, 288 pp.
ISBN 90-277-2639-6

M.J. Vishik and A.V. Fursikov: *Mathematical Problems in Statistical Hydromechanics*. 1988, 588 pp. ISBN 90-277-2336-2

V.I. Fabrikant: *Applications of Potential Theory in Mechanics. A Selection of New Results*. 1989, 484 pp. ISBN 0-7923-0173-0

E. Tirapegui and D. Villarroel (eds.): *Instabilities and Nonequilibrium Structures (II). Dynamical Systems and Instabilities*. 1989, 328 pp. ISBN 0-7923-0144-7

V.G. Bagrov and D.M. Gitman: *Exact Solutions of Relativistic Wave Equations*. 1990, 344 pp. ISBN 0-7923-0215-X

Other *Mathematics and Its Applications* titles of interest:

O.I. Zavialov: *Renormalized Quantum Field Theory.* 1990, 560 pp.
ISBN 90-277-2758-9

S.S. Horuzhy: *Introduction to Algebraic Quantum Field Theory.* 1990, 360 pp.
ISBN 90-277-2722-8

V.G. Makhankov: *Soliton Phenomenology.* 1990, 461 pp.   ISBN 90-277-2830-5

S. Albeverio, PH. Blanchard and L. Streit: *Stochastic Processes and their Applications in Mathematics and Physics.* 1990, 416 pp.   ISBN 0-9023-0894-8

V.A. Malyshev and R.A. Minlos: *Gibbs Random Fields. Cluster Expansions.* 1991, 248 pp.   ISBN 0-7923-0232-X

E. Tirapegui and W. Zeller (eds.): *Proceedings of the 3rd Workshop on Instabilities and Nonequilib ium Structures.* 1991, 370 pp.   ISBN 0-7923-1153-1

C.A. Marinov and P. Neittaanmaki: *Mathematical Models in Electrical Circuits. Theory and Applications.* 1991, 160 pp.   ISBN 0-7923-1155-8

C. Bartocci, U. Bruzzo and D. Hernandez-Ruiperez: *The Geometry of Supermanifolds.* 1991, 242 pp.   ISBN 0-7923-1440-9

E. Goles and S. Martinez: *Statistical Physics, Automata Networks and Dynamical Systems.* 1992, 208 pp.   ISBN 0-7923-1595-2

A. van der Burgh and J. Simonis (eds.): *Topics in Engineering Mathematics.* 1992, 266 pp.   ISBN 0-7923-2005-3

L. Aizenberg: *Carleman's Formulas in Complex Analysis.* 1993, 294 pp.
ISBN 0-7923-2121-9

E. Tirapegui and W. Zeller (eds.): *Instabilities and Nonequilibrium Structures IV.* 1993, 371 pp.   ISBN 0-7923-2503-6

G.M. Dixon: *Division Algebras: Octonions, Quaternions, Complex Numbers and the Algebraic Design of Physics.* 1994, 246 pp.   ISBN 0-7923-2890-6

A. Khrennikov: p-*Adic Valued Distributions in Mathematical Physics.* 1994, 270 pp.   ISBN 0-7923-3172-9

Z. Haba, W. Cegła and L. Jakóbczyk (eds.): *Stochasticity and Quantum Chaos.* (Proceedings of the 3rd Max Born Symposium, Sobótka Castle, September 15–17, 1993) 1995, 224 pp.   ISBN 0-7923-3230-X